APPLICATIONS OF GENERALIZED NETS

APPLICATIONS OF GENERALIZED NETS

Editor: Krassimir T Atanassov

Institute for Microsystems, Sofia, Bulgaria

World Scientific
Singapore • New Jersey • London • Hong Kong

Published by

World Scientific Publishing Co Pte Ltd
5 Toh Tuck Link, Singapore 596224
USA office: 27 Warren Street, Suite 401-402, Hackensack, NJ 07601
UK office: 57 Shelton Street, Covent Garden, London WC2H 9HE

British Library Cataloguing-in-Publication Data
A catalogue record for this book is available from the British Library.

APPLICATIONS OF GENERALIZED NETS

ISBN-13 978-981-02-0667-3
ISBN-10 981-02-0667-4

To Prof. Carl-Adam Petri on the occasion of the 30-th anniversary of his beautiful idea named Petri net.

PREFACE

This book contains papers on almost all models of processes developed up to now, which are described on the basis of Generalized Nets (GNs). These extensions of Petri nets were described in my book "Generalized Nets" (World Scientific Publ. Co., Singapore, 1991). The results used from the book "Generalized Nets" are given in Appendix 1.

This book is divided in five areas: theoretical results on GNs, theoretical GN-models, GN-models of economical, industrial and transportational processes, GN-models of processes, related to medical diagnostics, and GN-models of computer processes. Elements of the first version of the Program Package for GNs (PPGN), which is already prepared are described in the last chapter. The authors are my colleagues, postgraduate students and students, who are interested in GNs. The aim of this book is to show the largest possible scope of applications of GNs. I hope that this task will be fulfilled. GNs are realized in the frames of PPGN. Those who are interested in them can receive detailed information from me home address: Bulgaria, Sofia-1113, ul. Elemag, No. 15, Block 307, Ap. 18.

*
* *

I would like to thank the Chief Editor of AMSE Review (France) Prof G. Mesnard for urging me to write this book and my coauthors for the effort.

Krassimir T. Atanassov
April 1992, Sofia

CONTENTS

Chapter 1:

NEW RESULTS IN THE THEORY OF GENERALIZED NETS

Three new results related to the theory of GNs are included in this chapter. They are continuations of the results from the book "Generalized Nets".

§ 1.1: NEW RESULTS IN THE THEORY OF GENERALIZED NETS

Krassimir T. Atanassov

New results in the theory of GNs, obtained in the first five mounts of 1991, are introduced below. They are continuations of these from my book (noted here by [*]; see Appendix 1). A part of the results are based on [1, 2].

*
* *

§ 1.1.1: METHODOLOGICAL ASPECT OF THE THEORY OF GENERALIZED NETS

Various mathematical tools are used for the development and extension of the theory of GNs. Algebraic, logical, topological and other aspects of the theory of GNs were introduced during the last ten years. GNs are used as tools for describing some mathematical problems and some real processes. GN's application is possible in the following areas:

(a) tools for modelling, simulation and forecast of the processes,

(b) real-time control and/or optimization of non-critical in time real-time processes.

Figure 1.1 represents the general scheme of the GN's applications. When the real process modelled by a GN is complex enough, it is possible to use a combination of two or all the three functions of the GN-model.

After the construction of a given GN-model, questions about its validity and adequacy are important.

In this case, it is necessary to make a comparison with the reality. To this aim we need some validity criteria. Users modelling with GNs must define such criteria, which give them "enough evidence" for their results. There exist different methods for reaching this aim.

An idea for such a criterion may be as follows:

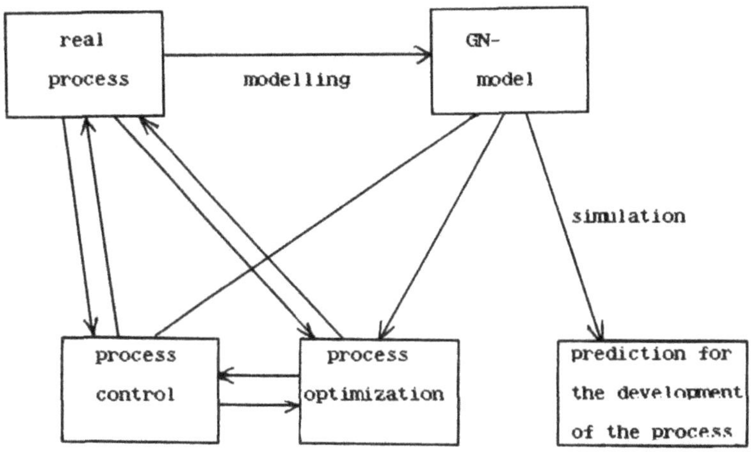

Fig. 1.1

Let the sets $K_1, K_2, \ldots, K_s \subset K(E)$ for $s \in N$, be randomly chosen; let, for every i $(1 \le i \le s)$, the sets

$$X_i = \{x_0^\alpha / \alpha \in K_i\} \subset X(K_i)$$

be randomly chosen also, and let

$$Y_i = \{x_{fin}^\alpha / \alpha \in K_i\},$$

where x_0^α and x_{fin}^α are the initial and final characteristics of the token α, respectively. Let the function $F: (P(K(E)))^s \to P(K(E))$ be such that

$$F(K_1, K_2, \ldots, K_s) = K \subset \bigcup_{i=1}^{s} K_i \subset K(E),$$

where P(A) is the set of all subsets of the set A. The above property of the function F can be considered in a way as an "averaging property". The sets

$$X = \{x_0^\alpha \;/\; \alpha \in K\}$$

and

$$Y = \{x_{fin}^\alpha \;/\; \alpha \in K\}$$

can be constructed directly and let

$$\bar{Y} = \{x_{fin}^\alpha \;/\; (\exists i:\; 1 \le i \le s)(\alpha \in K_i \;\&\; x_{fin}^\alpha \in Y_i)\}.$$

Let $\beta(Y, \bar{Y})$ be a criterion for "closeness" between the sets Y and $\bar{Y}$, which is defined by

$$\beta(Y, \bar{Y}) = \begin{cases} 1, & \text{if the sets } Y \text{ and } \bar{Y} \text{ are "close" (in some sense)}, \\ 0, & \text{otherwise.} \end{cases}$$

If for a number s large enough, and the equality

$$\beta(Y, \bar{Y}) = 1$$

is valid, we say that GN E is statistically significant (in the other case - it is non-statistically significant). If a given GN E is statistically significant, we can assert that every result of its functioning over a set of tokens $K' \subset K(E)$ with a set of initial characteristics $X' \subset X(E)$ is statistically significant.

There are other ways of constructing this criterion.

In GN-modelling, it is possible that the adaptive method with the form from Fig. 1.2 is the most positive one. In this case, the corrections in the constructed GN-model are done adaptively on the basis of the difference between the expected and obtained results (in relation with the same fixed criteria). We must bear in mind that the constructed GN-model is realized by the same program (e.g. in the framework of a "Program Package for GNs" PPGN, now being developed as shown in Fig. 1.2; any other program tool which realises GNs can take the place of "PPGN"). Some errors can occur in this program and corrections may be necessary, but this is not in the framework of the present description. Thus, we do not discuss the possibility of corrections in the program tools for GN realizations. However, there exists a probability to realise this too, if the results show existing defects in the programs.

As shown in Fig. 1.2, the process is simulated many times and the results are compared (and eventually averaged), analysed and estimated.

An analogy of the above scheme can be used when research is carried out with Petri nets or their extensions. But, for example, algebraic operations over GNs and hierarchical operators over GNs (see § 5 and § 8 in App. 1) give possibility of more complex models. These GN-models can be constructed by two main types:

- union,
- hierarchical.

To describe different processes which function in parallel, we can construct GN-models for each one of them and thus construct the GN which is a union of all these GN-models.
After this we proceed as in the above (simple) case. Now, the corrections of the model are made in the general GN (see Fig. 1.3).

There are two basic ways of constructing the general GN in the frames of the union types:

- simultaneous,
- sequential.

In the case of the simultaneous union type, the different (elementary) GNs are constructed preliminarily and the new (general) GN is obtained as a union of them. The validity of the new GN is checked as shown in Fig. 1.2 and the corrections are made in the general GN (see Fig. 1.3).

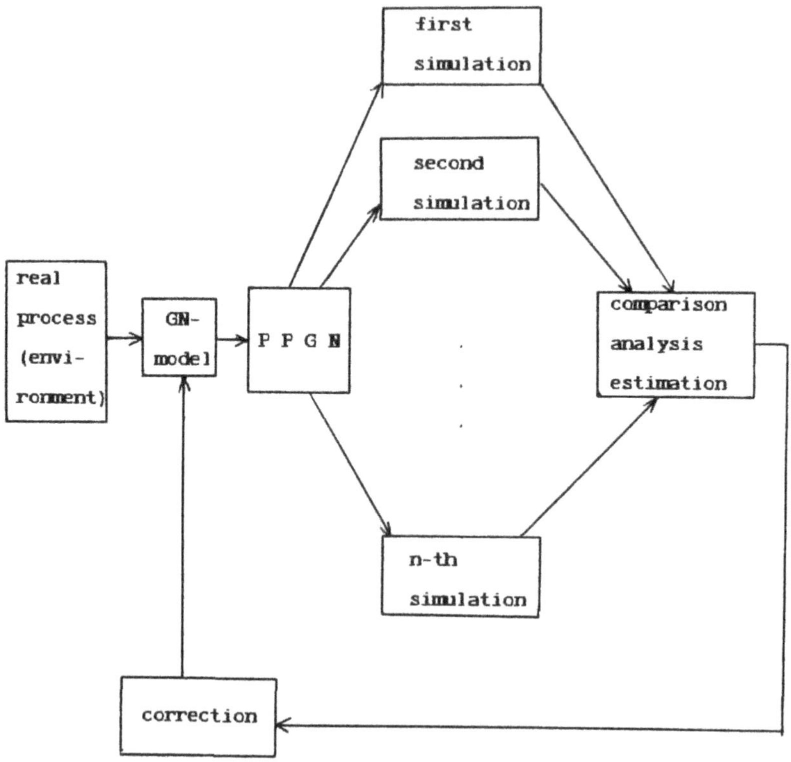

Fig. 1.2

In the case of the sequential union type, we give priorities to the different (elementary) GNs, and we start constructing the new (general) GN as a union of the current particular general GN and the GN with a current number. Initially, the particular general GN is an empty GN and the GN with a current number is the GN with the highest priority; finally, the particular general GN is the same general GN and the GN with current number is the GN with the lovest priority).

In every step of the last process of GN-constructing, we make a check for the validity of the functioning of the particular general GN and, if necessary, make corrections in the GN with the current number, because the other GN has already been tested. Thus, the number of tests is larger, but the corrections are made in simpler nets.

The other extension of the scheme from Fig. 1.2 (when the GN is obtained by hierarchical type ways) is shown in Fig. 1.4. In this case, there exist the same two types:

- simultaneous,
- sequential,

and the constructions are similar to the above ones. The main difference between the hierarchical type sequential way and the union type sequential way is as follows: in the second case, the growing of the current particular general GN is on the basis of nets with equal level, while in the first case the growth is in two directions: growing of the levels and growing of the subnets on the same level (i.e. places or transitions of the basic GN are changed with new subnets, in which it is made the same, etc.: places or transitions of a subnet of the GN from the last level are changed with new subnets).

Note, that there is no a restriction on the growth of the current particular general GN on levels and in the level, i.e. sequentially, different new subnets on the same level, as well as different subnets from different levels, can be added to it.

Nets, which describe some processes like the process of the solution of a transportation problem (see § 3.7 [*]), the process of the solution of a travelling salesman problem (see § 2.2) and others can be included in the frames of the addition to a fixed GN. These GNs are tested preliminarily as independent GNs and can be added to different GN-models. In future, we shall construct GNs which describe some sche-

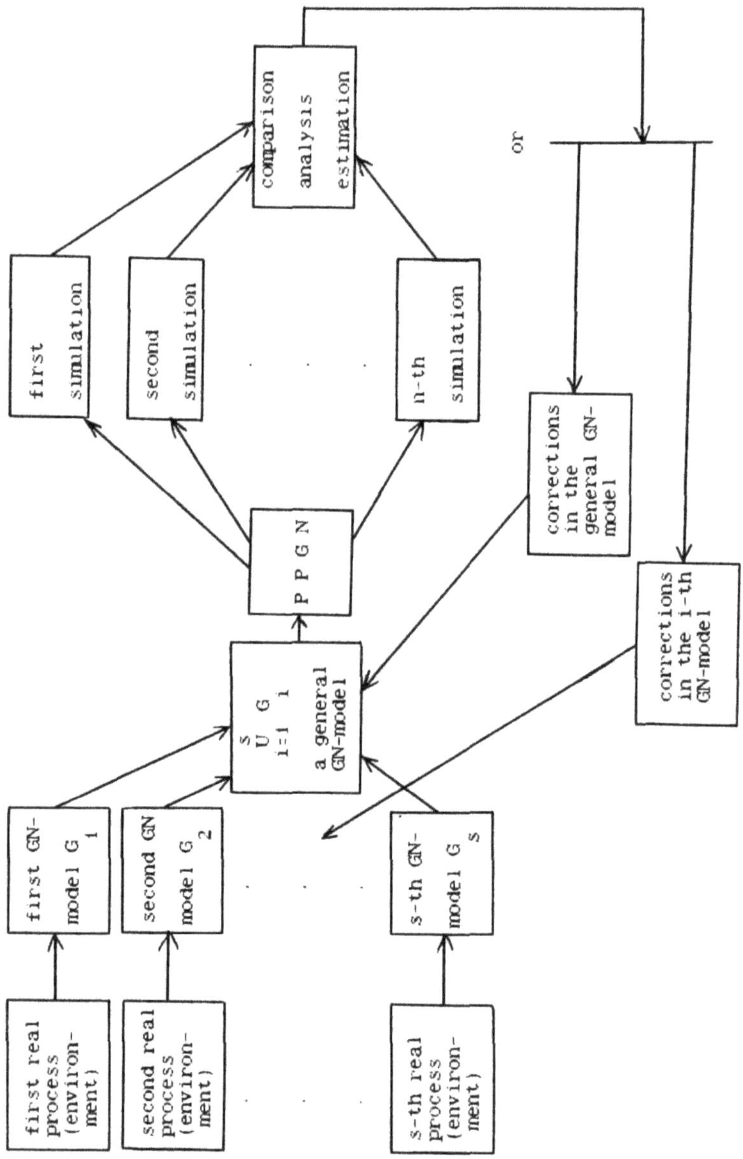
first real process (environment)
second real process (environment)
s-th real process (environment)
first GN-model G 1
second GN model G 2
s-th GN-model G s
s U G i=1 i
a general GN-model
P P G N
first simulation
second simulation
n-th simulation
comparison analysis estimation
or
corrections in the general GN-model
corrections in the i-th GN-model

Fig. 1.3

duling problems, other modifications of the two mathematical problems mentioned above and other optimization problems.

The above schemes illustrate one of the ways of using GN-models: GN-simulation of the modelled processes.

The confirmation of the validity can be obtained after more simulations of the process with the GN-model and after comparing the results. Thus we can say that the constructed GN-model is statistically reliable.

With these remarks, we shall briefly describe the process of GN-model's construction. Then, the GN-model can be used, as a tool for simulation, control and optimization (of the modelled real processes). We shall show some approaches of using GNs for these aims.

The first of these approaches is the sequential. Initially, on the basis of the GN-model, the corresponding real process can be controlled, e.g. with the help of an expert or a dispatcher. This is possible (see § 8 in App. 1) by virtue of the check on the suitable token transfer strategy. Different data related to the modelled process are accumulated (as token's characteristics and as data in the global GN-memory) in the GN-model in the framework of the process control. The accumulated information (from the necessary number of seances) is analysed, estimated and manipulated by statistical means and on the basis of these, the values of some parameters of the random functions related to different transition condition predicates and the values of different characteristic functions (which correspond to some GN-places) are determined. On this basis we can construct a new GN-model (with the same graphical structure) of the process (the alternation will be reached only with the help of a suitable dynamical operator) which can simulate the modelled process.

A variant of this procedure is as follows. Initially, the values of the parameters in the random functions which are in the places

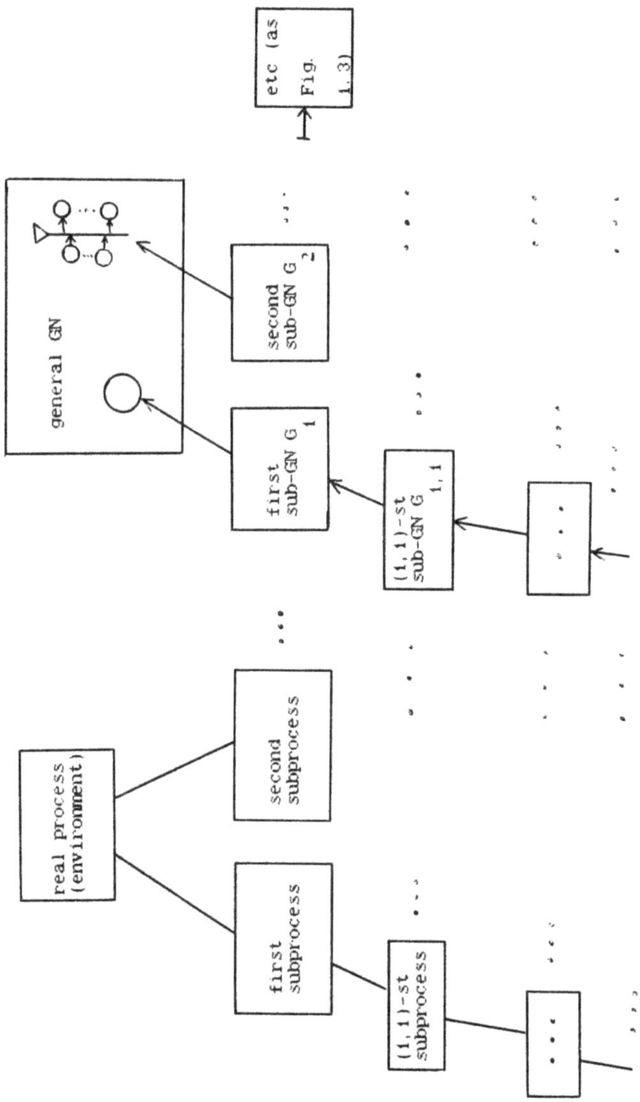
real process (environment)
first subprocess
second subprocess
(1,1)-st subprocess
general GN
first sub-GN G 1
second sub-GN G 2
(1,1)-st sub-GN G 1,1
etc (as Fig. 1.3)

Fig. 1.4

of some transition condition predicates are determined by an expert. The results of the simulation process are compared and, if necessary, these values are changed with others. After this, the control of the real process starts and the above procedure is repeated.

In both cases the process of precising the random values of the transition condition predicates and the place's characteristic functions can be an adaptive one (as in Figs. 1.2 and 1.3).

This situation, although more complex, will be valid when the GN-model is also used as an optimization tool.

To illustrate the above reasonings we shall discuss the possibility of using GNs as a tool for describing of global processes. There are two approaches: ascending and downgrading.

The ascending approach is based on the above discussed procedures for the GN-model constructions. For example, if we have to model the functioning of some chemical works, we can initially construct the GN-models of the different work sections, and later construct the GN which corresponds to the work in general. In this net the token will receive as current characteristics those values which the tokens of the GNs corresponding to the different work sections will receive as the final characteristics; and each of these sections will be represented in the global GN by a certain place or a certain transition.

The downgrading approach is the opposite. Initially, we construct the GN, which represents the basic sections of the work and the relation between them (the general GN) and later we start to construct the GNs which correspond to the different works' sections. In this case we must use the expert information about the works' processes when we determine the transition condition predicates and the place characteristic functions. In a way, the places of the general GN are "black boxes".

Finally, we shall discuss the possibility to use the GNs for the formulation of mathematical problems, and more precisely (here) for the generalization of such problems.

We shall use the first GN from § 2.2 as the first example. We see that in this GN we do not use all the GN's possibilities for modelling. Actually, the token which represents the travelling salesman can get some characteristics related to the travelling salesman as a person. He (the man, represented as a token) can carry different types of products in corresponding quantities, prices, limits of fitness. The built-up areas where the travelling salesman can also come can be specified by the nesessary quantities of products for them, distances between places and costs of transport between them, if there exists a relation between them and others. The number of travelling salesmen can be more than one (as in some generalizations of this problem) but now we can discuss the problems in which the travelling salesmen are representatives of one or more firms, i.e., they are or are not in competition among themselves. In § 2.2, a GN which describes a travelling salesman problem in some of these new modifications is discussed (the second GN).

Our second example will be related to the concept "expert system", which is a basic element of the new and modern scientific area "artificial intelligence". In a series of papers [3-6] we construct GNs which describe the functioning and results of the work of every expert system from a given class expert systems. Three tokens with initial characteristics: "hypothesis", "data base" and "list of rules" enter the most complex and general one of these GNs, which are described in [6]. The first token is split during the process of a GN-transfer into several different tokens and each of them leaves the net with a characteristic "the hypothesis is valid" or "the hypothesis is not valid". In the first case, the second token receives as the next characteristic the initial characteristic of the first token and the process continues (at a given moment other tokens with other hypotheses can enter the net). The third token is not modified during the time of GN-functioning. This is the real situation in the expert systems too. On the other hand, there exists some research related to the addition of new rules to the knowledge base and this process can also be described by a certain GN. Thus we can construct a new GN which describes both processes when they function in parallel. This is not realized in the expert systems yet.

Moreover, the token of the third type from the new GN can receive a current characteristic in relation to the values of the tokens of the first two types. Thus, on the basis of GN-models, self-modification expert systems can be constructed. For example, time-components can also be added to these expert systems. This idea will be the object of another research. The generalization of the neural networks, introduced in [7] on the basis of [8] (cf. § 2.1) is the third example of the application of GNs as a tool for the generation of extensions of existing processes, problems and concepts. The idea of this extension of the neural networks (in the frames of a given neural network, new neurons are generated with connections between them and some other neurons) is introduced after comparing the GN-components used in the GN from [8] to the other un-used GN-possibilities.

We can produce in an analogous way some other GN-models too.

GN-interpretations of some ideas from [9-15] and other books will be discussed in a next methodological paper on GNs.

*
* *

§ 1.1.2: MORE ON THE CONCEPT "GENERALIZED NETS"

One more definition of the concept GN will be introduced here. All other definitions of this concept are very complex. Now, based on the concept reduced GN, we shall discuss the properties of an object E, having the form

$$E = \langle A,\ K,\ X,\ \Phi\rangle,$$

where

(a) A is a set of transitions, each one of them having the form

$$Z = \langle L',\ L'',\ r\rangle,$$

where

L' and L" are the sets of the input and output transition places,

r is the transition condition;

(b) K is the set of tokens;

(c) X is the set of token's initial characteristics, (following § 2 from App. 1, we can also delete the component X, but in this case the description will be complicated),

(d) Φ is the token's characteristic function.

All GNs' elements stated above are as within the definition of ordinary GNs.

Note that this definition, similar to the ordinary GN definition, is not fully formalized. If we fully formalize the transition conditions and characteristic functions of the GNs, a smaller class of GNs will be obtained.

Obviously, a GN defined in such a way is reduced about the ordinary elements of the class Σ of all GNs. On the other hand (cf. § 2 in App. 1), each ordinary GN can be represented by a certain GN with the above form. Therefore the above GN has the modelling possibilities of the other types of GNs. There are processes which can be described by simpler types of GNs and in some cases, the reduced GNs described above are suitable as tools for modelling.

Below we shall discuss the basic properties of these reduced GNs, which we can call GNs, without restriction. Let Σ' be the class of all such GNs. From this point of view the ordinary GNs can be accepted as conservative extensions of the reduced one and $\Sigma \vdash \Sigma'$.

The algorithms for token's transfers in the frames of a fixed transition (for the abstract type of transitions also) and for the functioning of a fixed ordinary GN can also be used for the reduced GN (without essential corrections).

By analogy with ordinary GNs, we can construct fuzzy and intuitionistic fuzzy (reduced) GNs from both types, colour GNs, GNs with interval time for activation, GNs with a complex structure, GNs with an optimization component, GNs with a global memory, GNs with an additional clock.

It is important to note, that all other Petri net modifications are represented by the (reduced) GNs, having almost the same form as the ordinary GN.

We shall discuss in detail the operations and relations over Σ'. We shall repeat the definitions from § 5 on App. 1 with the corresponding changes.

For the two transitions Z_1 and Z_2 of a fixed GN E, we shall define

$$Z_1 = Z_2 \ \underline{\text{iff}}\ (\forall i: 1 \le i \le 3)(pr_i Z_1 = pr_i Z_2);$$

$$Z_1 \subset Z_2 \ \underline{\text{iff}}\ (\forall i: 1 \le i \le 2)(pr_i Z_1 \subset pr_i Z_2) \ \&\ (pr_3 Z_1 \bar{\subset} pr_3 Z_2),$$

where

$\bar{\subset}$ is a relation of inclusion over index matrices and if $A = [K_1, L_1, \{a_{i,j}\}]$, $B = [K_2, L_2, \{b_{i,j}\}]$, then

$$A \bar{\subset} B \ \underline{\text{iff}}\ (K_1 \subset K_2) \ \&\ (L_1 \subset L_2) \ \&\ (\forall i \in K_1)(\forall j \in L_1)(a_{i,j} = b_{i,j}).$$

Over the transitions $Z_i = \langle L'_i, L''_i, r_i \rangle$ $(i = 1, 2)$, we shall define three operations. It is necessary for the following to be valid: if place $l \in pr_1 Z_i \cap pr_2 Z_i$ and $l \in pr_s Z_{3-i}$, then $l \in pr_{3-s} Z_{3-i}$ for $1 \le i \le 2$, $1 \le s \le 2$. These operations are

(a) a union (the necessary condition for this operation is: if place $l \in pr_s Z_i$, then it is not possible, that $l \in pr_{3-s} Z_{3-i}$ for $1 \leq i \leq 2$, $1 \leq s \leq 2$)

$$Z_1 \cup Z_2 = \langle L_1^1 \cup L_1^2, L_2^1 \cup L_2^2, r^1 + r^2 \rangle;$$

(b) an intersection (with the above conditions)

$$Z_1 \cap Z_2 = \langle L_1^1 \cap L_1^2, L_2^1 \cap L_2^2, r^1 \times r^2 \rangle;$$

(c) a composition (with the above condition and with the condition

$$L_1^1 \cap L_1^2 = L_2^1 \cap L_2^2 = \emptyset),$$

$$Z_1 \circ Z_2 = \langle L_1^1 \cup (L_1^2 - L_2^1), L_2^2 \cup (L_2^1 - L_1^2), r^1 \circ r^2 \rangle .$$

As in § 5 in App. 1, we shall formulate some conditions which the arguments of these operations (different GNs) must satisfy.

We shall assume that if a place takes part simultaneously in two nets, then in both locations the place has

(a) the same characteristic functions,

(b) equal possibility of accepting tokens from K, i.e. if in the first net (which is a model of a certain process) the tokens, which are elements of the set $K' \subset K$ can pass through the examined place, and in the second net the tokens which are elements of the set $K'' \subset K$ can pass through this place, then $K' = K''$,

(c) one and the same numeration/notation.

(d) equal data in these tokens of both nets, which contain informations about both places (i.e. equal values of the place's priority function and of the place's capacity function).

Similarly, we shall expect that if two places are respectively an input one and an output one for two transitions in both nets, then the tokens of both nets, which contain informations about

(a) the capacities of the connecting arcs,

(b) the time for passing from the input to the output place, must have equal data. Also, these places should be in the same consequence, i.e. the place which is an input (output) one in the first of the transitions should be the same in the other transition too.

Let E_1 and E_2 be two GNs and for $i = 1, 2$ let:

$$E_i = \langle A_i, K_i, X_i, \Phi_i \rangle.$$

A union will be called the object

$$E_1 \cup E_2 = \langle A_1 \bar{\cup} A_2, K_1 \cup K_2, X_1 \cup X_2, \Phi_1 \cup \Phi_2 \rangle,$$

where

$$A_1 \bar{\cup} A_2 = \bigcup_{i=1}^{2} \{Z/(Z \in A_i) \ \& \ (\forall Z' \in A_{3-i})(Z \cap Z' = Z^{o})\} \cup$$

$$\bigcup_{i=1}^{2} \{Z/(\exists Z' \in A_i)(\exists Z'' \in A_{3-i})(Z' \cap Z'' \neq Z^{o}) \ \& \ (Z = Z' \cup Z'')\}.$$

Here the operations "union" and "composition", defined over ordinary GNs, coincide with the above GNs.

Local operator O and operation "iteration" in § 5 in App. 1 are defined in the same way.

For each two GNs E_1 and E_2, having the above forms,

$$E_1 = E_2 \ \underline{\text{iff}} \ (\forall i \ 1 \le i \le 5)(pr_i E_1 = pr_i E_2),$$

$$E_1 \subset_* E_2 \ \underline{\text{iff}} \ (\forall Z_1 \in A_1)(\exists Z_2 \in A_2)(Z_1 \subset Z_2) \ \& \ (K_1 \subset K_2)$$

$$\& \ (X_1 \subset X_2) \ \& \ (\Phi_1 = \Phi_2 | E_1),$$

where $g = h|E_1$ denotes that the function g is a restriction of the function h over the first GN E_1. Let for both functions g and h with

$Dom(g) = Dom(h)$:

$$g \le h \text{ \underline{iff} } (\forall x \in Dom(g))(g(x) = h(x)).$$

$E_1 =_o E_2$ <u>iff</u> $(\forall i\ 1 \le i \le 3)(pr_i E_1 = pr_i E_2)$;

$E_1 {}^I\!\subset_o E_2$ <u>iff</u> $(A_1 \subset A_2)$ & $(\emptyset \ne Q^I_1 \subset Q^I_2)$ & $(K_1 \subset K_2)$,

$E_1 \subset^O_o E_2$ <u>iff</u> $(A_1 \subset A_2)$ & $(\emptyset \ne Q^O_1 \subset Q^O_2)$ & $(K_1 \subset K_2)$,

the relation $E_1 \bar{\subset}_o E_2$ is not defined here,

$E_1 \subset_o E_2$ <u>iff</u> $(E_1 {}^I\!\subset_o E_2)$ v $(E_1 \subset^O_o E_2)$.

$E_1 \subset_U E_2$ <u>iff</u> $(A_1 \subset A_2)$ & $(K_1 \subset K_2)$ & $(X_1 \subset X_2)$ & $(\Phi_1 = \Phi_2 | E_1)$,

For a given token $\alpha \in K$ and a given initial characteristic $x \in X$ let

$$E(\alpha, x) = \begin{cases} \langle \alpha, x^{\alpha}_{fin} \rangle, & \text{if } \alpha \in K \text{ and } x \in X \\ \langle \alpha, x \rangle, & \text{otherwise} \end{cases},$$

$$E\{\alpha, x\} = \begin{cases} \langle \alpha, x, x^{\alpha}_1, x^{\alpha}_2, \ldots, x^{\alpha}_{fin} \rangle, & \text{if } \alpha \in K \text{ and } x \in X \\ \langle \alpha, x \rangle, & \text{otherwise} \end{cases},$$

where x^{α}_{fin} is the final characteristic of the token α in the GN E.

Let us define

$$E_{\emptyset}(\alpha, x) = E_{\emptyset}\{\alpha, x\} = \langle \alpha, x \rangle.$$

Below we shall introduce four other relations between GNs:

$E_1 \sqsubset E_2$ <u>iff</u> $(K_1 \subset K_2)$ & $(X_1 \subset X_2)$ & $(\forall \alpha \in K_1)(\forall x \in X_1)$

$$(E_1(\alpha, x) = E_2(\alpha, x)),$$

$$E_1 \sqsubset_* E_2 \ \underline{\text{iff}}\ (K_1 \subset K_2) \ \& \ (X_1 \subset X_2) \ \& \ (\forall \alpha \in K_1)$$

$$(\forall x \in X_1)(\exists i_1, i_2, \ldots, i_s : 1 \le i_1 < \ldots < i_s \le \text{fin})$$

$$(E_1\{\alpha, x\} = pr_{i_1, i_2, \ldots, i_s} E_2\{\alpha, x\}),$$

$$E_1 \approx E_2 \ \underline{\text{iff}}\ (E_1 \sqsubset E_2) \ \& \ (E_2 \sqsubset E_1),$$

$$E_1 \approx_* E_2 \ \underline{\text{iff}}\ (E_1 \sqsubset_* E_2) \ \& \ (E_2 \sqsubset_* E_1).$$

Obviously, the relations $\sqsubset_*$ and $\approx_*$ are stronger than the relations $\sqsubset$ and $\approx$, respectively.

The examples and assertions from Chapter 5 [*] are valid without essential corrections, but the proofs are (obviously) simpler. The theorem for the completeness of the GN's transitions is also valid.

The complexity operators $\varphi_1, \varphi_2, \varphi_3, \varphi_5, \varphi_7, \varphi_8, \ldots, \varphi_{13}$ are applied to the (reduced) GN without change. This is valid for all graph-operators too.

All assertions from Chapter 6 [*] are valid here without essential corrections (cf. § 6 in App. 1).

The same is valid for non-temporal logical operators from Chapter 7 [*].

The problem with the application of different types of operators is more complex. Some of the global ones, as G_1, G_2, G_3, G_4, G_6, G_{11} and G_{13} can be applied directly. Others are applied only "indirectly", i.e., in the following way: when a corresponding (ordinary) GN's component (which is represented in the (reduced) GN with the characteristic of a certain special token) must be changed by some of these operators, this can be done by operator G_{12}. Local operators P_5,

P_8, P_9, P_{10} are applied directly and the others are represented by operator G_{12}. The hierarchical operators are modified for the present case by the means discussed above.

The reduced operator is defined only over the set $\{A_1, A_2, A_5, f, K, X, \Phi\}$ and, as in the case with the ordinary GNs, the sets Σ^{A_1}, Σ^{A_2}, Σ^{A}, Σ^{K} are empty.

All extending operators (with necessary additions) can be defined over the (reduced) GN too. Moreover, we can now define an extending operator, which will juxtapose a new ordinary GN to every (reduced) GN. As shown above, this extension is conservative.

When the information about arc's capacities is determined in some token's initial characteristics, all $A_{1,j}$-type dynamical operators ($1 \leq j \leq 18$) can be applied over the (reduced) GN. On the other hand, the operators $A_{1,j}$ are applied only where $j \in \{2, 5, 8\}$. The operators of $A_{i,j}$-type where $2 \leq i \leq 4$ and j is arbirtarily admissible, are applied directly.

Based the above definitions we can construct self-modifying reduced GNs and universal reduced GNs.

Finally, we shall emphassize that GNs are simpler than ordinary ones, but more processes exist whose models by ordinary GNs are much more convenient.

* * *

§ 1. 1. 3: UNIVERSAL GENERALIZED NETS. II.

We shall discuss two problems related to the concepts of Universal GNs (UGNs) and Self-Modifying GNs (SMGNs) (see § 9 in App. 1) and to the GN-problems from § 11 in App. 1:

1. the existence of a Universal GN (UGN) for the class Σ of all GNs,
2. the existence of a UGN for the class Σ_{SMGN} of all SMGNs.

In § 9. 1 [*] we have shown that if a fixed UGN had started functioning before a given GN, then the UGN may not to be able to represent the functioning of the given GN. However, the proof of Theorem 1 [17] is wrong. Following Theorem 2 [17] and Theorem 1 of § 9 in App. 1, we construct a UGN for which

(a) $\bar{T} = \max\{T \ / \ (\exists E \in \Sigma)(T = pr_1 pr_3 E)\}$

(it is possible: $\bar{T} = -\infty$);

(b) $\bar{t}^{o} = \min\{t^{o} \ / \ (\exists E \in \Sigma)(\bar{t}^{o} = pr_2 pr_3 E)\}$

(everywhere $\bar{t}^{o} > 0$) and

(c) $\bar{t}^{*} = \max\{t^{*} \ / \ (\exists E \in \Sigma)(\bar{t}^{*} = pr_3 pr_3 E)\}$

(it is possible: $\bar{t}^{*} = +\infty$);

then for each GN, the GN from Theorem 2 [17] and Theorem 1 of § 9 in App. 1 will be a UGN, i. e. (this is the author's mistake in [17]) the $(-\infty)$-UGN of the class $\Sigma_{-\infty}$ is a UGN for the class Σ. This GN will function at every moment and its tokens will be as the tokens of the UGN from Theorem 2 [17]. Therefore the following is valid:

<u>THEOREM 1. 1. 3. 1</u>: There exists a GN which is a UGN for the class Σ.

All other results from [17] are valid. Hence for every subclass of GNs, a UGN exists which represents the functioning and results of the work of each element of this subclass.

THEOREM 1.1.3.2: There exists a GN, which is an UGN for Σ_{SMGN}.

Proof: Let the GN E^* from Fig. 1.5 be given and

$$r_1 = \begin{array}{c|ccc} & l_4 & l_5 & l_{16} \\ \hline l_1 & W_1 & false & W_2 \\ l_4 & W_3 & W_4 & false \end{array}$$

where

W_1 = "the token is α"

(We assume that card(K) > 0 because in the opposite case, the GN E is not valid.);

$W_2 = \neg W_1$;

W_3 = "$c(l_1, TIME) > 1$";

$W_4 = \neg W_3$;

$$\square_1 = \wedge(l_1, l_4);$$

$$r_2 = \begin{array}{c|c} & l_6 \\ \hline l_5 & W_5 \\ l_{11} & W_5 \end{array}$$

where

W_5 = "$(\exists Z \in pr_1 pr_1 x^{\beta}_{cu})(pr_3 Z = x^{\gamma}_{cu})$",

$$\square_2 = \vee(l_5, l_{11});$$

$r_3 =$

	l_7	l_8	l_9
l_6	true	false	false
l_2	false	W_6	W_7
l_{12}	false	W_6	W_7

where

W_6 = "in the last characteristic of the token from l_6 there is at least one GN-control token"

(This is possible because the token α has the highest priority.);

$W_7 = \neg W_6$;

$$\square_3 = \wedge(l_6, \vee(l_2, l_{12}));$$

$r_4 =$

	l_{10}	l_{11}	l_{12}	l_{13}	l_{14}	l_{15}
l_7	W_7	W_8	false	false	false	false
l_8	false	false	W_8	W_7	false	false
l_9	false	false	W_8	W_7	false	false
l_3	false	false	false	false	W_7	W_8
l_{15}	false	false	false	false	W_7	W_8

,

where

W_8 = "$x_{cu}^{\Gamma} < pr_1 pr_3 x_0^{\beta} + pr_3 pr_3 x_0^{\beta}$",

$W_7 = \neg W_8$,

$$\square_4 = \wedge(l_7, \vee(l_8, l_9), \vee(l_3, l_{15}));$$

$$r_5 = \begin{array}{c|cc} & l_{17} & l_{18} \\ \hline l_{10} & W_9 & W_{10} \\ l_{16} & false & true \end{array},$$

where

W_9 = "$\bar{c}(l_{16}$, TIME) > 0"

$W_{10} = \neg W_9$,

$$\square_5 = \wedge(\vee(l_{10}, l_{17}), l_{16}).$$

Let an arbitrary SMGN E with a set of tokens K be given, where

$$K = \{\alpha_1, \alpha_2, \ldots, \alpha_n\}.$$

Let the tokens of E enter place l_1 of E^* (with the E-characteristics, i.e. with their initial E-characteristics). Let the token α with the maximum possible priority (e.g. $\pi_K(\alpha) = \infty$) be also in l_1 without initial characteristic. Token β with an initial characteristic: "the description of E" enters place l_2 and the token Γ with an initial characteristic $pr_1 pr_3 E$ (= T - the initial time of the functioning of E) enters place l_3. All E-tokens from l_1 enter sequentially in place l_{16} without characteristics. Token α enters place l_4 and there it receives sequentially as a characteristic: "the characteristics of all E-tokens and their priorities" (in their priority-order). After that, the token α enters place l_5 without a characteristic. The E-tokens wait for the end of the process of functioning of E^* in place l_{16}.

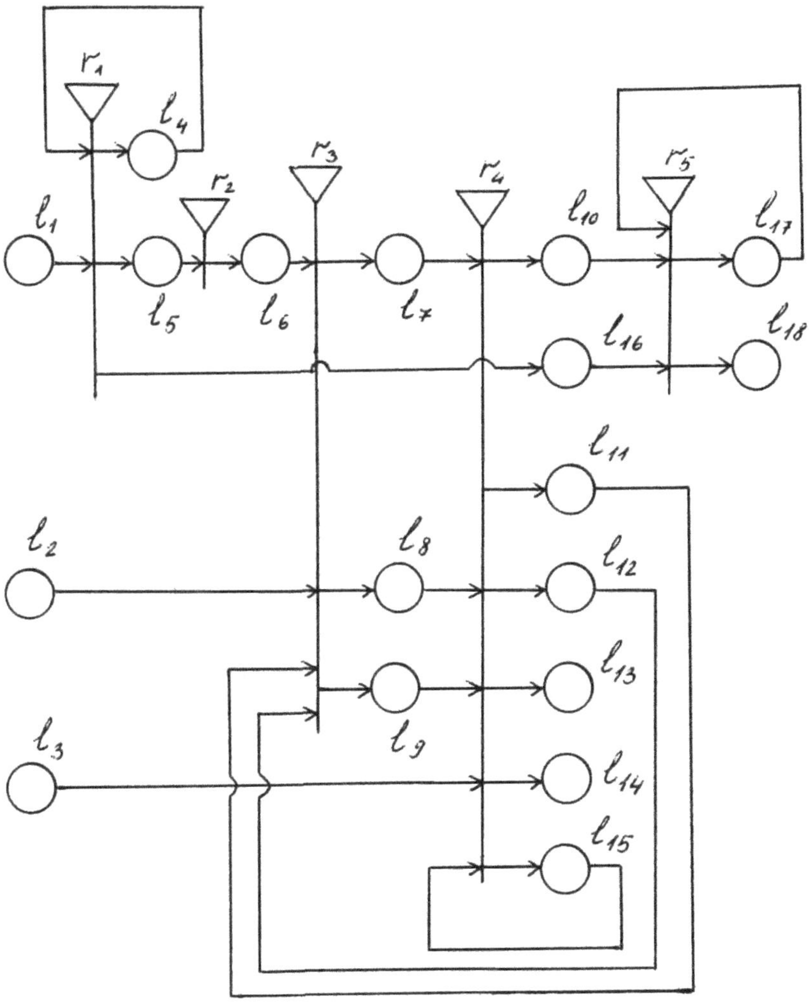
r1
l4
r3
r2
r4
r5
l1
l10
l17
l5
l6
l7
l16
l18
l11
l2
l8
l12
l13
l9
l3
l14
l15

Fig. 1.5

There they will receive their final characteristics. The priorities of the three tokens (α, β and Γ) satisfy the condition

$$\pi_K(\alpha) > \pi_K(\beta) > \pi_K(\Gamma).$$

The token α enters place l_6 with a list of these E-tokens which are transferred in the current time-moment from input to output E-transition places and a list of the corresponding characteristics of these E-tokens. For a more detailed description of this process, the place l_6 must be changed (by the hierarchical operator H_1 with a GN from the type of the GN from Theorems 9.1.1 of § 9.1 in [*]. We do not describe precisely this element of the net E^* because it will only complicate its description.

As a result of the activation of the transition Z_3, the tokens α and β enter places l_7 and l_8 or l_9, respectively. In place l_7 the token α will not receive a characteristic. The token β will enter place l_8, if in the current characteristic of α there exists at least one special E-token; otherwise β enters place l_9. In the first case β receives as a new characteristic the new description of the SMGN E, whose description is received as a result of applying operator(s) over E (the type of this (these) operator(s) is (are) determined as a characteristic of the token α. In the second case, β does not receive a new characteristic.

Transition Z_4 is activated when the three tokens are in its input places. After this, token α enters place l_{10} or l_{11} without characteristics, token β enters place l_{12} or l_{13} without characteristics, token r enters place l_{14} or l_{15} without characteristic in l_{14} and with a characteristic "the new value of the current time" in l_{15}, in relatioin to the value of the current time.

Finally, transition Z_5 is activated when the time for the functioning of the SMGN E expires. The token α (which has a higher priority over the tokens from place l_{16}) enters place l_{17} without a characteristic. The tokens from l_{16} enter sequentially place l_{18} and there they receive as characteristic the list of all characteristics which are related to them. The list is received as a characteristic of the token α at the time of its transfer in the net E^*. Thus the E-tokens $\alpha_1, \ldots, \alpha_n$ will leave the GN E^* with the same characteristics as those tokens leaving the SMGN E.

Therefore the new net will represent the functioning of the SMGN E.

We shall denote for E^* as follows: all its places and transitions have equal priorities; the places l_1, l_{16} and l_{18} have infinite capacity; the others, capacities 1.

The GN time-components can be

(a) $T = 0$;

(b) $t^o = 1$;

(c) $t^* = \infty$;

(d) $\theta_1(t) = t + 1$;

(e) $\theta_2(t) = 1$;

(f) $\theta_K(\partial) = 0$

where $\partial \in \{\alpha, \beta, \Gamma\}$;

(g) $b(\alpha) = \infty$;

$b(\beta) = b(\Gamma) = 1$.

A GN constructed in such a way represents the functioning of each SMGN and therefore it is a UGN for Σ_{SMGN}.

Therefore Theorem 1.1.3.2 is proved and another proof of Theorem 1.1.3.1 is also given.

<u>THEOREM 1.1.3.3</u>: $\Sigma_{SMGN} \dashv\vdash \Sigma$, i.e. Σ_{SMGN} is a conservative extension of Σ.

Proof: From Theorem 1.1.3.2, it follows that each SMGN can be represented by the UGN, which is an element of Σ. Therefore every element of Σ_{SMGN} can be represented by an element of Σ and hence $\Sigma \vdash \Sigma_{SMGN}$. The other direction is obvious.

Thus we give an answer of the Problem 19 in App. 1.

REFERENCES:

[1] K. Atanassov Methodological aspect of the theory of generalized nets. II, AMSE Review Vol. 20, No. 4, 1992, 53-64.

[2] K. Atanassov, Universal generalized nets. II, AMSE Review, Vol. 21 (1992), No. 1, 59-64.

[3] K. Atanassov, L. Atanassova, E. Dimitrov, G. Gargov, I. Kazalarski, M. Marinov and S. Petkov, Generalized nets and expert systems. Methods of Operations Research, Vol. 59., Proc. of the 12-th Symposium on Operations Research of the Gesellschaft fur Mathematik, Okonomie und Operations Research, 1987, Passau, Frankfurt a.M: Athenaeum, 1989, 301-310.

[4] K. Atanassov, L. Atanassova, E. Dimitrov, G. Gargov, I. Kazalarski, M. Marinov and S. Petkov, Generalized nets and expert systems. II, Proc of Int. Conf. "Networks Information Processing

Systems", Sofia, May, 1988, Vol. 2, 54-67.

[5] K. Atanassov, L. Atanassova, E. Dimitrov, G. Gargov, I. Kazalarski, M. Marinov, S. Petkov and M. Stefanova-Pavlova, Generalized nets and expert systems. III. Methods of Operations Research, Vol. 63. Proc. of the 14-th Symposium on Operations Research. Ulm, Sept. 1989, 417-423.

[6] K. Atanassov, L. Atanassova, E. Dimitrov, G. Gargov, I. Kazalarski, M. Marinov and S. Petkov, Generalized nets and expert systems. IV. Proc. of the XIX Spring Conf. of the Union of Bulg. Math., Sunny Beach, April 1990, 155-161.

[7] L. Hadjyisky and K. Atanassov, A generalized net, representing the elements of one neuron network set, AMSE Review Vol. 14, No. 4, 1990, 55-59.

[8] L. Hadjyisky L. and K. Atanassov K. Theorem for representation of the neuronal networks by generalized nets, AMSE Review Vol. 12, No. 3, 1990, 47-54.

[9] Handbook of operations research foundations and fundamentals (J. Moder and S. Elmaghraby, Eds.), Van Nostrand Reinhold Co., New York, 1978 (Vol. 1 and 2).

[10] R. Ackoff, The art of problem solving, A Wiley-Interscience Publication, New York, 1978.

[11] J. Kleijnen, Statistical techniques in simulation, M. Dekker INC, New York, 1974 (Volss. 1 and 2).

[12] P. Fishburn, Utility theory for decision making, J. Wiley & Sons, INC, New York, 1970.

[13] H. Wagner, Principles of operations research, Prectice-Hall, Inc. 1969.

[14] J. Kanter, Management-oriented management information systems, Prentice-Hall Inc., Englewood Clitts, New Jersay, 1977.

[15] R. Ackoff and M. Sasieni, Fundamentals of operations research, John Wiley and Sons, Inc., New York, 1968.

[16] Atanassov K., Open problems in the theory of generalized nets, Preprint IM-MFAIS-1-91, Sofia, 1991.

[17] Atanassov K., Universal generalized nets, AMSE Review, Vol. 7 (1988), No. 4, 29-35.

§ 1.2: ON THE REPRESENTATION OF GENERALIZED NET TRANSITIONS BY A FIXED SET OF TRANSITIONS

Pavlin Gyurov and Krassimir Atanassov

Let a fixed GN and a set S with elements - fixed graphical structures of transitions be given. We shall describe an algorithm for a representation of the transitions of the given GN E by the elements of the set S. Thus we shall solve the problem 12 of App. 1:

"To construct an algorithm which for every set of transition graphical structures (obviously they must satisfy some conditions) and for every given GN constructs a GN for which

(a) all its transitions have forms which are elements of the given set;

(b) both nets have equal tokens with equal initial characteristics, and as a result of their functioning, the equal tokens receive equal final characteristics."

The mentioned conditions for the form of transition can be as follows:

$$\sum_{Z \in S} |L'_Z| \geq |S| + 1$$

(i.e. it is necessary that at least one transition of S exists which contains at least two input places) and

$$\sum_{Z \in S} |L''_Z| \geq |S| + 1$$

(i.e. it is necessary that at least one transition of S exists which contains at least two output places), where $|X|$ is the cardinality of the set X.

Let the above two conditions be satisfied for the elements of the set $S = \{Z_1, Z_2, \ldots, Z_s\}$ and let the GN

$$E = \langle\langle A, \pi_A, \pi_L, c, f, \theta_1, \theta_2\rangle, \langle K, \pi_K, \theta_K\rangle, \langle T, t^o, t^*\rangle, \langle X, \Phi, b\rangle\rangle$$

be given, where the elements of the set A of the GN's transitions have the forms

$$Z = \langle L', L", t_1, t_2, r, M, \square\rangle,$$

(see App. 1) and let $r = [L', L", \|r_{\mu,\nu}\|]$ (see App. 2).

From the Theorem for the completeness of the GN transitions (see § 5 of App. 1), it follows that every GN can be represented by a union, a composition and an iteration of its transitions, and therefore it is sufficient to show a method for representation of an arbitrary GN's transition and this can also be transformed in all other transitions.

The new GN will have transitions with S-forms. It is convenient that this GN be a GN with a global memory (see § 3 in App. 1). Let this new component contain as parameters the lists of current capacities of the transition's arcs and of the numbers of the free location in the output places: for a transition Z these lists are: $\bar{M}_Z = [L'_Z, L"_Z, \|\bar{m}_{i,j}\|]$, where $\bar{m}_{i,j}$ are the current capacities of the transition's arcs between places l_i and l_j; and $\{\langle l, c(l)-\bar{c}(l,TIME)\rangle / l \in L"_Z\}$, where $\bar{c}$ is a function giving the number of tokens in place l at the current time-moment TIME.

The GN transitions can be ordered according to their priorities. Let a transition Z of GN E be given.

Our algorithm is as follows: By "//.../" we shall denote some comments.

A01. Order the input and output places of Z by their priorities, where the order in the first case is descending and in the second case in ascending series.

A02. Check the inequality

$$|pr_1 Z_i| \geq |pr_1 Z| \tag{1}$$

for every i ($1 \leq i \leq s$), where $pr_j X$ is the j-th projection of the n-dimensional set X (1 ≤ j ≤ n) and $S = \{Z_1, Z_2, \ldots, Z_s\}$

If the set of the S-transitions which satisfy (1) is empty, go to A07.

A03. Check the inequality

$$|pr_2 S_i| \geq |pr_2 Z| \tag{2}$$

for every i $(1 \leq i \leq s)$.

If the set of the S-transitions which satisfy (2) is empty, go to A06.

A04. Determine this S-transition Z^*, which satisfies (1) and (2), for which

$$|pr_1 Z^*| + |pr_2 Z^*|$$

is minimum.

A05. Mark the first $|pr_1 Z|$ input places of Z^* with the corresponding place's identifiers of the input places of Z and the first $|pr_2 Z|$ output places of Z^* with the corresponding place's identifiers of the output places of Z; mark the other places of Z^* with other (different) identifiers - these places are elements of the set $(L'_{Z^*} - L'_Z) \cup (L''_{Z^*} - L''_Z)$ and define

$$t_1^{Z^*} = t_1^Z,$$

$$t_2^{Z^*} = t_2^Z,$$

$r_{Z^*} = r_Z + \bar{r}$,

where $\bar{r} = [L'_{Z^*} - L'_Z,\ L''_{Z^*} - L''_Z,\ \|false\|]$ is a predicative index matrix,

$M_{Z^*} = M_Z + \bar{M}$,

where $\bar{M} = [L'_{Z^*} - L'_Z,\ L''_{Z^*} - L''_Z,\ \|\ 0\ \|]$ is a natural number index matrix,

$\square_{Z^*} = v(\square_Z,\ v(\{l\ /\ l \in L'_{Z^*} - L'_Z\}))$.

Go to A20.

//Obviously

$$!\square_{Z^*} \text{ iff } !\square_Z,$$

because the tokens do not enter in the places of the set $L'_{Z^*} - L'_Z$.

The method of checking the functioning of a transition constructed in such a way is similar to the one shown below and that is why it will be omitted here.//

A06. Mark the first $|pr_1 Z|$ input places of Z^* with the corresponding place's identifiers of the input places of Z, and the other input places of Z^* with other (different) identifiers - these places are elements of the set $(L'_{Z^*} - L'_Z)$. Go to A10.

A07. Check the inequality (2) and if it is not valid, go to A08, else mark the first $|pr_1 Z|$ output places of Z^* with the corresponding place's identifiers of the output places of Z, and the other output places of Z^* with other (different) identifiers - these places are elements of the set $(L''_{Z^*} - L''_Z)$. Go to A08.

A08. Let $Z_0 = Z$. Let $i = 1$.

A09. Determine this transition of S with minimal number of input pla- places whose number is greater than $|Z_{i-1}| - 1$ or, if such a transition does not exist, the S-transition with the maximum number of input places (let it be Z'_i). Construct a transition Z_i from a transition Z_{i-1} by removing its final $|pr_1 Z'_i| - 1$ from the number of places. Mark the first input place of Z'_i by y'_i (an identifier which is not found among the other GN's identifiers) and the other input places of Z'_i with the identifiers of the corresponding input places of Z_{i-1}. The final output place of the transition Z'_i is marked by y'_{i-1} (it is the first input place of the transition Z'_{i-1}, if it exists). Mark all other output places of Z'_i with other identifiers which are not found among the other GN's identifiers: $x'_{i,1}, x'_{i,2}, \ldots, x'_{i,|L'_{Z'_i}|-1}$. Define

$c(x'_{i,j}) = 0,$

$\pi_L(x'_{i,j}) = 0$ (for $1 \le j \le |L'_{Z'_i}| -1$),

$c(y'_{i-1}) = c(y'_i) = \infty,$

$$\pi_L(y'_i) = \max \{\pi_L(l) \;/\; (l \in pr_2 Z'_i) \;\&\; (l \neq y'_i)\}.$$

If the transition Z_i has no input places (i.e. if the second case mentioned above is valid), then finish the procedure, and if (2) is valid for Z, determine the priorities and capacities of the other input places of Z'-transitions, in relation to the corresponding priorities and capacities of the places of Z and go to A12, else go to A10; else i:= i + 1 and execute A09 again.

A10. Let Z_0 = Z. Let i = 1.

A11. Determine this transition of S which has the minimum number of output places whose number is greater than $|Z_{i-1}| - 1$ or, if such a transition does not exist, the S-transition with the maximum number of output places (let it be Z''_i). Construct a transition Z_i from a transition Z_{i-1} by removing its final $|pr_1 Z''_i| - 1$ for the number of places. Mark the first output place of Z''_i by y''_i (an identifier, which is not found among the other GN's identifiers) and the other output places of Z''_i - with the identifiers of the corresponding output places of Z_{i-1}. The final input place of the transition Z''_i, is marked by y''_{i-1} (the first output place of the transition Z''_{i-1}, if it exists). Mark all other input places of Z''_i with other identifiers which are not found among the other GN's identifiers: $x''_{i,1}, x''_{i,2}, \ldots, x''_{i,|L''_{Z''_i}|-1}$. Define

$$c(x''_{i,j}) = 0,$$

$\pi_L(x''_{i,j}) = 0$ (for $1 \le j \le |L''_{Z'_i}| - 1$),

$c(y''_{i-1}) = c(y''_i) = \infty$,

$\pi_L(y''_i) = \max\{\pi_L(l) / (l \in pr_2 Z'_i) \& (l \neq y'_i)\}$.

If the transition Z_i has no output places (i.e. if the second case mentioned above is valid), then finish the procedure, determine the priorities and capacities of the other output places of Z"-transitions, in relation to the corresponding priorities and capacities of the places of Z and go to A12; else i:= i + 1 and execute A11 again.

A12. Identify places y'_0 and y''_0 by the operator $\bar{O}$ (see § 5 in App. 1).

//See Fig. 1.6. Nore that by these 12 steps the graphical structures of the transitions which represent the given transition Z are determined.

Below we shall determine the other components of the above constructed transitions.//

A13. Let the number of the transitions of the first (left) group be n_1 and that of the second (right) group be n_2. Determine a new time-step t' for which

$$t' = \frac{t^o}{n_1 + n_2 + 1}.$$

//More generally,

$$n_1 = \max_{Y \in S} pr_1 Y,$$

$$n_2 = \max_{Y \in S} pr_2 Y,$$

and these values can be determined before the application of this algorithm.//

A14. Determine the following time-components

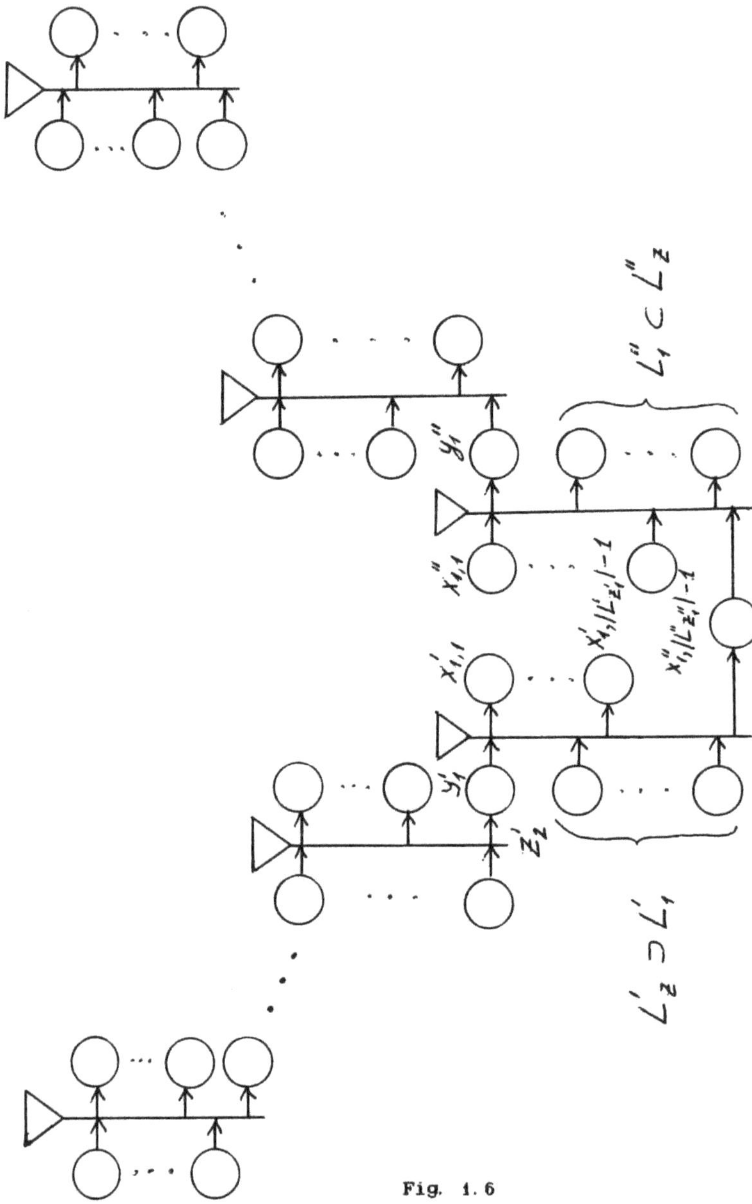

Fig. 1.6

$$t_1^{Z'_i} = t_1^{Z} + (n_1 - i).t' \quad (1 \le i \le n_1),$$

$$t_1^{Z''_i} = t_1^{Z} + (n_1 - 1 + i).t' \quad (1 \le i \le n_2).$$

A15. Determine all elements of the M-components of the transitions as "∞".

A16. Determine the transition types of all transitions as "v" (i.e. they will be activated when at least one token enters some of its places).

A17. Define the characteristic function $\bar{\Phi}$ in the different places as follows

$\bar{\Phi}_{y'_i}$ gives as the current characteristic of the token the list of the output Z-places in which the token must enter,

$\bar{\Phi}_{x'_{i,j}}$, $\bar{\Phi}_{x''_{i,j}}$, $\bar{\Phi}_{y''_i}$ are not defined,

$\bar{\Phi}_{l''_i}$ coincides with $\Phi_{l''_i}$ which is defined for the given GN.

A18. Determine the transition condition predicates of the i-th transition from the left group

$$
r_i = \begin{array}{l|ccccc}
 & x'_{i,1} & x'_{i,2} & \ldots & x'_{i,|L''_{Z'_i}|-1} & y'_{i-1} \\
\hline
y'_i & false & false & \ldots & false & true \\
l'_{i,1} & false & false & \ldots & false & W' \\
l'_{i,2} & false & false & \ldots & false & W' \\
\vdots & \vdots & \vdots & \vdots & \vdots & \vdots \\
l'_{i,|L'_{Z'_i}|-1} & false & false & \ldots & false & W'
\end{array}
$$

where

$$\bigcup_i \{l'_{i,1}, l'_{i,2}, \ldots, l'_{i,|L'_{Z'_i}|-1}\} = L'_Z,$$

and

$$W' = \text{"}(\exists\gamma: 1 \le \gamma \le |L'_{Z'_i}|)((f(r_{\mu,\gamma}) = 1)\ \&\ (\bar{m}_{\mu,\gamma} > 0)\ \&$$

$$(c(l''_\gamma) - \bar{c}(l''_\gamma, TIME) > 0))\text{"},$$

where l''_μ is the index of the μ-th row of the index matrix r_i.

//The values of $\bar{m}_{\mu,\gamma}$ and of $\bar{c}(l''_\gamma, TIME)$ are located in the global GN-memory.//

Determine the transition condition predicates of the j-th transition from the right group:

r_i =

	y'_i	$l''_{j,1}$	$l''_{j,2}$	...	$l''_{i,\|L''_{Z''_i}\|-1}$
$x'_{i,1}$	false	false	false	...	false
$x'_{i,2}$	false	false	false	...	false
⋮	⋮	⋮	⋮	⋮	⋮
$x'_{i,\|L'_{Z''_i}\|-1}$	false	false	false	...	false
y''_{i-1}	W	W"	W"	...	W"

where

$$\bigcup_j \{l''_{j,1}, l''_{j,2}, \ldots, l''_{j,|L''_{Z''_j}|-1}\} = L''_Z$$

and

W = "There exists at least one identifier of the place of a transition with a bigger number, which is an element of the set of the last token characteristic",

W" = "The identifier of the output place is an element of the set - the last token characteristic",

A19. Define: $\pi_A(Z'_i) = \pi_A(Z''_j) = \pi_A(Z)$ for every Z'_i and Z''_j - transitions from the left and from the right groups, respectively.

A20. Check the list of the transitions of GN E. If all transitions are processed - end of the procedure; else - go to A01.

Let the newly constructed GN be marked by E^*. From the above construction, it follows that all transitions of E^* are elements of the set S. Moreover, the last characteristics which the tokens will receive in Z''_j-transition places will coincide with the characteristics which the tokens will receive in the corresponding Z-places.

Therefore the following theorem is valid.

THEOREM: For every set of transition graphical structures, which satisfy the inequalities (1) and (2) and for every given GN, there exists a GN for which

- all its transitions have forms which are elements of the given set,
- both nets have equal tokens with equal initial characteristics and, as a result of their functioning, the equal tokens receive equal final characteristics.

This paper is based on [1].

REFERENCE:

[1] P. Gyurov and K. Atanassov, On the representation of generalized net transitions by fixed set of transitions, Preprint IM-MFAIS-2-91, Sofia, 1991.

§ 1.3: TWO NEW CONSERVATIVE EXTENSIONS OF GENERALIZED NETS

Rumen S. Christov and Krassimir T. Atanassov

Some problems related to the extensions of the concept "generalized net" (GN) were generated during the development of the program realization of GNs. Below we discuss two new types of nets and prove that they are conservative extensions of the ordinary GNs.

Let the object

$$<<A, \pi_A, \pi_L, c, f, \Theta_1, \Theta_2>, <K, \pi_K, \Theta_K>, <T, t^o, t^*>, <X, \Phi, b>; L'>$$

be given and let it have the following property: when in a certain place of the set $L' \subset L$ there exist at least two of its tokens, which are generated by one token "predecessor" independently from the fact that

(a) there can exist other "kindred" tokens somewhere (i.e. tokens with the same "predecessor"),

(b) the tokens can be generated in different moments and from different, but kindred, "predecessors",

then these tokens are united in one token by analogy with the procedure from § 8 in App. 1. The new token will also be a kindred token of the tokens which are kindred tokens of its generated ones.

Let us call this new net "a GN allowing a particular token's uniting" (GN-APTU).

<u>THEOREM 1.3.1</u>: $\Sigma_{GN\text{-}APTU} \subset \Sigma$.

<u>Proof</u>: Obviously, every GN-APTU for which $L' = \emptyset$, i.e. for which L' is not defined (L' = "*"), is an ordinary GN, i.e. $\Sigma_{GN\text{-}APTU} \vdash \Sigma$. Let a GN-APTU E be given. Initially, we construct an ordinary GN $\bar{E}$ with the same graphical structure. After this, for each output place $l \in L'$ of transition Z of GN-APTU E (let l be a fixed place), we construct a new transition Z_l which corresponds to l (this transition contains

place l as an input place). This new transition (see Fig. 1.7) has a transition condition with the form

$$r_1 = \begin{array}{c|ccc} & \bar{l} & l_1^Z & l_2^Z \\ \hline l & W_1 & W_2 & W_3 \\ & & & \\ l_2^Z & W_4 & false & W_5 \end{array},$$

where

W_1 = "the token has no kindred tokens in places l and l_2^Z";

W_2 = "in place l_2^Z there is already a kindred token of the present token";

$W_3 = \neg W_1 \;\&\; \neg W_2$,

W_4 = "the token has no kindred tokens in place l and it has received as a current characteristic the characteristics of the last token in the place l_1^Z" (this is described below in relation to the characteristic function),

$W_5 = \neg W_4$.

The following inequalities for the new places are valid

$$\pi_L(\bar{l}) > \pi_L(l_1^Z) > \pi_L(l_2^Z)$$

and

$$\pi_L(l) > \pi_L(l_2^Z).$$

Let the GN-characteristic function $\bar{\Phi}$ be defined in such a way that in place l_2^Z the token receives as new characteristic all the

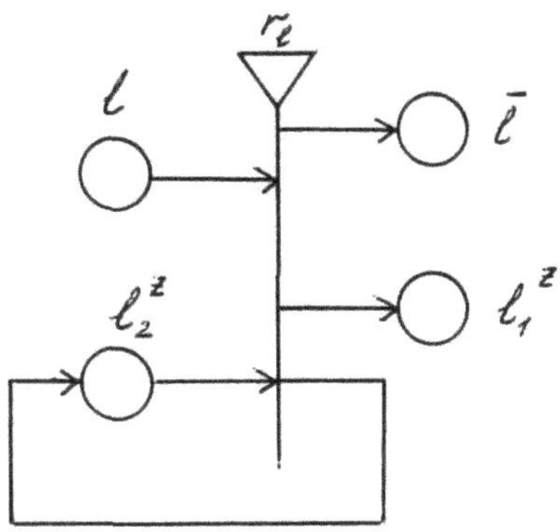

Fig. 1.7

characteristics, which the token, last entered place l_1^Z, has received after the time-moment when it has split with the token from place l_2^Z, or with its corresponding "predecessor". In place l_1^Z the tokens do not receive new characteristics. In place $\bar{l}$ the tokens receive the same characteristics as the tokens of GN-APTU E in place l.

For transition Z from E (we denote it by Z_E), let

$$L_Z = \{l/l \in L' \cap pr_2 Z_E\}.$$

Following the above algorithm we can construct the set of transitions of $\bar{E}$ $\{Z_l/l \in L_Z\}$, and to transition Z of E we juxtapose the transition

$$\bar{Z} = Z \cup \bigcup_{l \in L_Z} Z_l .$$

From this construction it is obvious that the transition Z from E will function in the same way as the transition $\bar{Z}$. By analogy with the proof of Theorem 5.3.1 [*] (see § 5 in App. 1) we can prove that the GNs E and $\bar{E}$ are equivalent in the results of their functioning. Therefore $\bar{E}$ represents GN-APTU E and hence $\Sigma \sqsubset \Sigma_{GN-APTU}$. ◊

The other extension of the concept GN is related to the characteristic functions of the nets. Now they are related to the tokens. Tokens receive new values from the characteristic functions associated with the places which the tokens enter. These functions may be dependent on the entire information in the net. On the other hand, the places are also influenced by the processes in the nets, but they do not have their characteristics. It can be obtained by defining the characteristic functions which will give characteristics to the places.

Let the net

$$E = \langle\langle A, \pi_A, \pi_L, c, f, \Theta_1, \Theta_2\rangle, \langle K, \pi_K, \Theta_K\rangle, \langle T, t^o, t^*\rangle, \langle X, \bar{X}, \Phi, \bar{\Phi}, b\rangle\rangle$$

be given, for which $\bar{\Phi}$ is a function determining the places' current characteristics. For this net the places have initial characteristics which are elements of the set $\bar{X}$. We shall call this net a GN with places' characteristic functions (GN-PCF).

The proof of the following assertion is easy.

<u>THEOREM 1.3.2</u>: $\Sigma_{GN-PCF} \sqsubset \Sigma$.

Every GN with a global memory (see § 3 in App. 1) can be represented by an ordinary GN. From the fact that all values of all characteristic functions (related to tokens or places) are calculated at different moments of the GN's functioning, it follows that these results can be remembered in the GN "memory" and therefore the GN "memory" can represent them. Therefore all GN-PCF can be represented by a GN with a global memory. On the other hand, this type of extensions of a GN can be represented by an ordinary GN, from which, the validity of the Theorem follows, because the opposite is obvious: every ordinary GN is a GN-PCF without the specific components. ◊

From the construction of the above proof follows the validity of

<u>THEOREM 1. 3. 3</u>: Every ordinary GN (or a GN-PCF) can be represented by a reduced GN with a global memory but without the set of initial characteristics and the characteristic functions.

Chapter 2:

SOME APPLICATIONS OF GENERALIZED NETS IN SCIENCE

In this chapter different results are included which are related to the application of the GNs in some areas of science such as applied mathematics, artificial intelligence and scientometrix. Thus illustrations of the possible applications of GNs in science are given. Four of the papers (§ 2.1, § 2.2, § 2.4 and § 2.6) are revised versions of old papers, and the other four contain new results.

§ 2.1: GENERALIZED NETS REPRESENTING THE ELEMENTS OF NEURON NETWORKS

Ljubomir Hadjyisky and Krassimir Atanassov

The class of Neuron Networks (NNs) (see [1-3]) which will be represented with GNs is a variety of the "additive model" [4]. This class of NNs is described by the equation

$$Y(t + \partial t) = g(I(t)),$$

where

(a) $I(t) = T.Y(t)$;

(b) N is the number of the neurons in the network;

(c) t is the time-parameter;

(d) $Y(t) = \langle y_1(t), y_2(t), \ldots, y_N(t)\rangle^T \in R^N$

(R is the set of all real numbers, $\langle \ldots \rangle^T$ is the transposed vector of the indicated one.) is the state of the NN at the time-moment t;

(e) $y_i(t)$ is the state of the neuron "i" at the time-moment t;

(f) $T \in R^{N \times N}$ is a matrix whose elements $T_{i,j}$ $(1 \le i, j \le N)$ are the synaptic strength of the connection between the neuron "j" and the neuron "i";

(g) $$I(t) = \langle I_1(t), I_2(t), \ldots, I_N(t)\rangle^T \in R^N$$

is the potential of the somma obtained by integration of the separate input impacts;

(i) $$I_i(t) = \sum_{j=1}^{N} T_{i,j} . y_j(t)$$

$(1 \le i \le N)$;

(j) $g(I(t)) = \langle g_1(I(t)), g_2(I(t)), \ldots, g_N(I(t))\rangle^T \in R^N$

is an $(N \times 1)$-vector function determining the state of the neurons.

(R^N is a linear N-dimensional vector space, $R^{N \times N}$ is a linear (NxN) -dimensional vector space.)

$y_i(t)$ describes the state of the neuron "i". This state can be exited or not and, according to this $y_i(t)$, accepts correspondingly the values 1 or 0.

The determination of $g_i(I_i(t))$ is realised by the following relations:

$$y_i(t) = g_i(I_i(t)) = \begin{cases} 1 & , \text{ if } I_i(t) > \Theta_i \\ 0 & , \text{ if } I_i(t) < \Theta_i \\ y_i(t), & \text{ if } I_i(t) = \Theta_i \end{cases},$$

where Θ_i is the threshold value for the neuron "i".

This class of NNs has the possibility of self-training, realised by changes of the synaptic streingths $T_{i,j}$ or the threshold values Θ_i. In case some neurons are dead, the network can reorganize itself and the other neurons will accept the functions of the dead ones. The representation with GNs includes this property as well.

Let the GN E be composed of two graphical incoherent GNs, E_1 and E_2 (see Fig. 2.1). GN E_1 is a modelled somatical potential $I(t)$ and a determined state of the neurons $Y(t + \partial t)$ by $g(I(t))$. GN E_2 performs self-training and reorganization when a neuron is dead. The GN E_1 is composed of nine places, three transitions and N tokens, where N is the number of neurons in the network. The capacity of every place l_i is $c(l_i) = N$ for $1 \leq i \leq 9$, i.e. at some time-moment all tokens can fall into one of the places. It is convenient to use an extended GN with a global memory (see § 3 in App. 1). In this case, in the com-

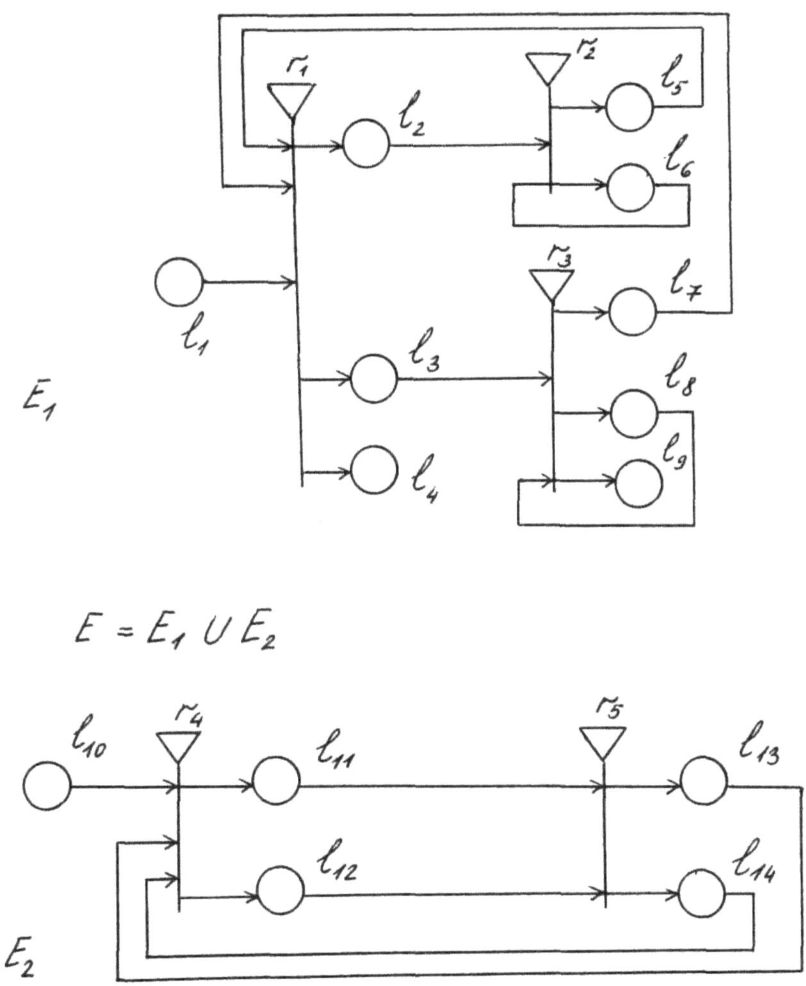

Fig. 2.1

ponent B different data sets about the net-topology $\{T_{i,j}\}$, the somatical potential of neurons $\{I_i(t)\}$, their threshold values $\{\Theta_i\}$, their state $\{y_i(t)\}$ can be disposed of as follows

(a) $$\sum_{i=1}^{N}{}^{+} \begin{array}{c|cccc} & J_i(1) & J_i(2) & \dots & J_i(r_i) \\ \hline i & T_{J_i(1),i} & T_{J_i(2),i} & \dots & T_{J_i(r_i),i} \end{array},$$

where $\sum_{i=1}^{N}{}^{+}$ marks the sum of N index matrices (see App. 2), i is the number of the neuron "i", r_i is the number of the neurons which are connected with neuron "i", $\{J_i(k) \,/\, 1 \le k \le r_i\}$ is the set of the numbers of the neurons which are connected with the neuron "i", for $1 \le \mu \le r_i$, $T_{J_i(\mu),i}$ is the synaptic strength of μ-th connections,

(b) $$\sum_{i=1}^{N}{}^{+} \begin{array}{c|c} & i \\ \hline i & I(t) \end{array},$$

(c) $$\sum_{i=1}^{N}{}^{+} \begin{array}{c|c} & i \\ \hline i & \Theta_i \end{array},$$

(d) $$\sum_{i=1}^{N}{}^{+} \begin{array}{c|c} & i \\ \hline i & y_i(t) \end{array}.$$

The relation among places is given by the conditions of the index matrices for the corresponding transitions:

$$r_1 = \begin{array}{c|ccc} & l_2 & l_3 & l_4 \\ \hline l_1 & W_1 & W_2 & W_3 \\ l_5 & W_1 & W_2 & W_3 \\ l_7 & W_1 & W_2 & W_3 \end{array},$$

where

W_1 = "$y_i(t) = 1$" & $\neg W_3$,

W_2 = "$y_i(t) = 0$" & $\neg W_3$,

W_3 = "$\gamma_\alpha = 0$";

$$r_2 = \begin{array}{c|cc} & l_5 & l_6 \\ \hline l_2 & W_4 & W_5 \\ l_6 & W_4 & W_5 \end{array},$$

where

$W_4 = W_2 \vee W_3$,

$W_5 = W_1$;

$$r_3 = \begin{array}{c|ccc} & l_7 & l_8 & l_9 \\ \hline l_3 & W_6 & W_7 & W_8 \\ l_9 & W_6 & W_7 & W_9 \end{array},$$

where

$W_6 = W_1 \vee W_3$,

$W_7 = W_2$,

W_8 = "no connection exists",

w_9 = "$C = 0$",

where

$$\gamma_\alpha = V_\alpha(t) = \begin{cases} 1, & \text{if a neuron } \alpha \text{ is a live,} \\ 0, & \text{if a neuron } \alpha \text{ is a dead,} \end{cases}$$

$V_\alpha(t)$ is the function determining the physical existence of the neuron α.

The characteristic functions Φ_i $(1 \le i \le 9)$ are not defined. Note, that if we use an ordinary GN, i.e. a GN without global memory B, then a part of the characteristic functions will exist. Hence the function Φ is not defined.

The time-moments for firing and for activity duration of the transitions are

$$t_1^1 = t_o,$$

$$t_1^2 = t',$$

$$t_1^3 = t_o,$$

$$t_2^1 = \infty,$$

$$t_2^2 = t'',$$

$$t_2^3 = \infty,$$

where

t_o (= T) is the initial time-moment of the NN and hence of the GN E,

$$t' = \begin{cases} t_o + t_n & \text{, if } t = t_o \text{ and there are no tokens in the} \\ & \quad \text{place } l_6, \\ t_{calc} + t_{E_2} + t'', & \text{otherwise,} \end{cases}$$

t_{calc} is the time at which $\{I_i(t)\}$ and $\{y_i(t + \partial t)\}$ will be calculated,

t_{E_2} is the time at which the unique token of the GN E_2 does one circle in E_2,

t'' is the time at which N tokens transfer from input places to output places of the transition Z_2,

t_n is the time at which a token can be transferred from the transition Z_1 to the transition Z_2.

The calculation of $\{I_i(t)\}$ and $\{y_i(t + \partial t)\}$ begins at the time-moment t'. When $t = t_o$, the values $\{y_i(t)\}$ of some neurons which are inputs in the NN can be changed and this will be an external impact to the NN.

The GN E_2 is composed of five places, two transitions and one token. The capacity of every place is 1. In this GN, self-training and reorganization are accomplished. Bearing in mind that the previous states of the NN are remembered as a matrix $\{y_i^s\}$ $(1 \le s \le n)$, where n is the number of this state. They are selected by determined criteria in the component B of the GN E.

The relations among the places are given by the conditions:

$$r_4 = \begin{array}{c|cc} & l_{11} & l_{12} \\ \hline l_{10} & W_{10} & W_{11} \\ l_{13} & W_{10} & W_{11} \\ l_{14} & W_{10} & W_{11} \end{array}$$

where

W_{10} = "R = 1",

$W_{11} = \neg W_{10}$,

$$r_5 = \begin{array}{c|cc} & l_{13} & l_{14} \\ \hline l_{13} & W_{12} & W_{13} \\ l_{14} & W_{12} & W_{13} \end{array}$$

where

W_{12} = "P = 1",

$W_{13} = \neg W_{12}$,

where the parameters R and P have the following values:

$$R = \begin{cases} 1, & \text{if } C_{d,old} = C_d, \\ 0, & \text{if } C_{d,old} < C_d, \end{cases}$$

where $C_{d,old}$ is the number of dead neurons before the last cycle, C_d is the current number of dead neurons;

$$P = \begin{cases} 1, & \text{if non-self-training}, \\ 0, & \text{if self-training}. \end{cases}$$

The values $C_{d,old}$, C_d and P are defined as parameters for the component B of the GN E. The value of P is defined at the initial time-moment. The characteristic functions Φ_i $(10 \le i \le 14)$ are not defined. The time-moments for firing and activity duration of the transitions Z_4 and Z_5 are

$$t_1^4 = t_{calc},$$

$$t_2^4 = t_{tr},$$

$$t_1^5 = t_o \ (= T),$$

$$t_2^5 = t_{tr},$$

where

t_{tr} is the time when one token transfers from an input place to an output place of one transition,

$t_{E_2} = t_{r_4} + t_{r_5}$,

$$t_{r_4} = \begin{cases} t_{tr}, & \text{if } R = 1, \\ t_{E_2}^{reor}, & \text{if } R = 0, \end{cases}$$

$$t_{r_5} = \begin{cases} t_{tr}, & \text{if } P = 1, \\ t_{E_2}^{self}, & \text{if } P = 0, \end{cases}$$

t_{r_4} is the time when a token is transferred throughout the transition Z_4,

t_{r_5} is the time when a token is transferred throughout the transition Z_5,

$t_{E_2}^{reor}$ is the time when the NN reorganizes when a neuron is dead,

$t_{E_2}^{self}$ is the time when the NN is self-trained.

The calculation of new characteristics of the NN when the network has been reorganized begins at the time-moment t_{calc} and when it has been self-trained, at the time-moment $t_{calc} + t_{r_4}$.

Let T be the starting moment at which GN E begins functioning, t be the running time of the model, given by a time-variable, different from the time-variables of the GN E, defined for the component B of the GN.

At the moment T, all tokens are in the place l_1. The transition Z_1 is fired ($t_1^1 = T$, $t_2^1 = \infty$) and the tokens of l_1 are determined in the places l_2, l_3 and l_4. The tokens which are disposed in place l_2 symbolize the neurons which are excited and which are not dead; in place l_3, which are not excited (inhibiting or normal state), in place l_4, which are dead. When the neurons fall into the place l_4, they stay there and do not influence the functioning of the NN. Neurons of l_3 are transferred in places l_7 and l_8 (transition Z_3 is fired). Neurons which change their state from non-excited to excited or dead fall in place l_7; in place l_8, neurons which keep their state and are alive. The conditions for tokens transfer from l_8 in l_7, or from l_8 again in l_8, are the same as the ones from l_3 to l_8 or l_7. Only excited neurons can change the state of other neurons from the network. In this case tokens (neurons) transfer through the transition Z_2. The change in the neuron state will be described in detail.

At the starting-moment, when $t = T = t_o$, there are tokens in place l_6; the transition Z_2 is fired at the time-moment $t_o + t_n$, i.e. when a token appears in place l_2 and the duration of the fired transition is t". At the time-moment $t_o + t''$ the calculation of $\{I_i(t)\}$, $\{y_i(t)\}$ and E_2 begins as follows: let a token with a number α be transferred through Z_2. The somatical potentials $I_{J_\alpha(k)}$ of the neurons connected with neuron α will be

$$I_{J_\alpha(k)}(t) = T_{J_\alpha(k),\alpha} . Y_\alpha(t),$$

with $k = 1, 2, \ldots, r_\alpha$.

$I_{J_\alpha(k)}(t)$ is the change of the potential of neuron $J_\alpha(k)$ and it is added to the old value of this potential defined as an element of the component B. In the same way, for all other tokens passing through Z_2, the potentials of the neurons are determined in the NN. When this calculation is finished for all tokens, potentials I_α $(1 \le \alpha \le N)$ are determined completely. Then the calculation continues with the function $g_\alpha(I_\alpha(t))$ for $1 \le \alpha \le N$ and the new states $\{y_i(t + \partial t)\}$ are determined. The running time t is equal to t_{calc}. At this moment the transition Z_4 of E_2 is fired and the token moves to place l_{11} or l_{12}. If the token falls in place l_{11}, a reorganization is not necessary, i. e. the number of neurons does not decrease. If a token falls in place l_{12}, a reorganization is necessary in the data of the neurons in the NN and the connections among them. At the time

$$t = t_{calc} + t_{r_4} + t'',$$

the reorganization is completed and the transition Z_5 is fired. The token falls in place l_{14} or l_{13}, where the corresponding self-training is either performed or not at the time t_{r_5}. When GN E_2 is working, transitions Z_1 and Z_3 are fired and the excited tokens (neurons) are moved correspondingly from place l_8 to place l_7, and l_2. The dead neurons are moved correspondingly from place l_8 to place l_7 and l_4. When the running time t is equal to the following:

$$t = t_{calc} + t_{E_2} + t'',$$

the transition Z_2 is fired and the running time t receives the value t_0. So excited tokens are removed from places l_2 and l_6 to place l_6; the non excited and dead ones are entered in to place l_5. In this way the process continues. If all tokens fall in place l_8 this means that the not excited neurons in the network are absent and all tokens are entered into place l_9 and the GN E stops functioning.

By the representation of NNs with GNs, the distribution of neurons by their functional state is obtained. This allows a better visualization during the NN functioning, because only excited neurons take part in the corresponding operations. On the other hand, it is possible that some of the operations for the neuron's data change are performed simultaneously.

These properties obtained by the representation with GN as a possibility for an external impact of neuron characteristics made possible the effective modelling with NNs of real processes connected with processing of information.

*
* *

Below we shall define a new type of NNs and construct a GN E^* which will represent their functioning and the results of their work.

Each new NN has the same elements (neurons) as the NNs from above. The functions, which determine the functioning, the state and the potential (Y, I, g, etc.) of the new NNs are the same as the first NNs, described by the GN above.

The basic difference between the two types of networks is the possibility of new neurons being born in the second type of NNs. These neurons possess the same functional possibilities as the others in the network. In the second type of NNs new functions exist (which are defined in the global memory B of the GN E^*):

(a) $V_1(t)$, which determines the number of the neurons born at the time-moment t;

(b) $V_2(t) = \{\alpha_1, \alpha_2, \ldots, \alpha_{V_1(t)}\}$ - the set of born neurons;

(c) $V_3(N, t) = N + V_1(t)$ - the new neuron number for NN (N is the number of neurons in the NN before time-moment t;

(d) $V_4(x, t) = T_{x,j} \cup T_{i,x}$, where $1 \le i, j, x \le V_3(N, t)$, $x \in V_2(t)$ and $T_{x,j}$, $T_{i,x}$ are as above;

(e) $V_5(x, t) = \theta_x$ for $x \in V_2(t)$ (for θ_x see above);

(f) $V_6(x, t) = g_x(I_x(t))$ for $x \in V_2(t)$ (for g_x and I_x see above).

The functions V_1 and V_4 are stochastic or they can be related to the network parameters. The functions V_2, V_3, V_5 and V_6 are strongly connected with V_1 and its type.

At time-moment t of being borning, every new neuron $x \in V_2(t)$

has a zero potential $(I_x(t) = 0)$ and a zero state $(Y_x(t) = 0)$, where Y_x is defined by g_x and I_x as from above.

Now the function $g(I(t)) \in R^{\Omega}$, where Ω is the number of the natural numbers.

The GN E^* has the form of Fig. 2.2 (cf. Fig. 2.1). All notations which are identical in both nets E^* and E are marked identically. The new places in E^* are m_1, m_2, m_3 and m_4; the new transitions are Z_6 and Z_7 (a transition Z_5 of GN E does not exist here). The transition Z_7 realizes the connection between the subnets E^*_1 and E^*_2, which are analogous to E_1 and E_2 in the above GN E.

The transition conditions in the GN E^* are as follows:

$r_1 =$

	l_2	l_3	l_4
l_1	W_1	W_2	W_3
l_5	W_1	W_2	W_3
l_6	W_1	W_2	W_3
m_4	W_1	W_2	W_3

where

W_1 = "y_i(TIME) = 1" & $\neg V_3$;

W_2 = "y_i(TIME) = 0" & $\neg V_3$;

W_3 = "$\theta_\alpha = 0$" (for θ_α see above).

The transition conditions r_2, r_3 and r_4 are the same as above.

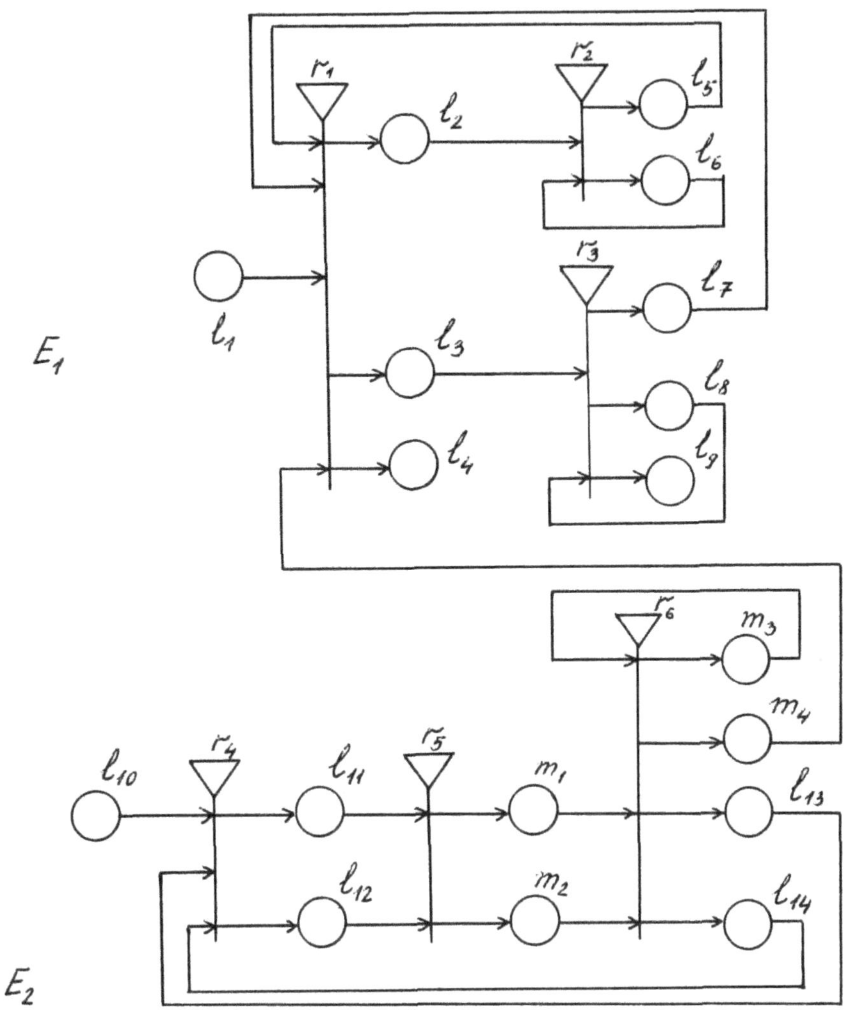

Fig. 2.2

The transition Z_5 does not exist and hence the same is valid for r_5.

$$r_6 = \begin{array}{c|cc} & m_1 & m_2 \\ \hline l_{11} & W_{14} & W_{15} \\ l_{12} & W_{14} & W_{15} \end{array}$$

where

W_{14} = "V_1(TIME) > 0";

W_{15} = "V_1(TIME) = 0".

In place m_1 the control token (of the second subnet of GN E^* which corresponds to the GN E_2 from above), receives a characteristic "V_1(TIME)", where TIME is the current time-moment of the GN-model. The condition

$$r_7 = \begin{array}{c|cccc} & m_3 & m_4 & l_{13} & l_{14} \\ \hline m_1 & W_{16} & W_{16} & W_{17} & W_{18} \\ m_2 & W_{16} & W_{16} & W_{17} & W_{18} \\ m_3 & W_{16} & W_{16} & W_{17} & W_{18} \end{array}$$

where

W_{16} = "x_{last} > 0";

W_{17} = "x_{last} = 0" & W_{12};

W_{18} = "x_{last} = 0" & W_{13}

(W_{12} and W_{13} are defined above).

In place m_3 the token receives as characteristic "x_{last} - 1" and in place m_4: "<x, V_4(x, TIME), V_5(x, TIME), V_6(x, TIME), I_x(TIME),

Y_x(TIME)>", where $x \in V_2$(TIME).

All components in the GN global memory are changed to new ones for which N:= N + V_1(TIME) at the time-moment TIME.

In this GN, the boundary of N is Ω and N is a function of the global model-time. Here the numbers of new connections between neurons are added to the elements r_i, if they are generated for the corresponding neurons.

In this GN

$$t_{E_2} = t_{r_4} + t_{r_6} + t_{r_7},$$

where t_{r_4} is defined above,

$$t_{r_6} = \begin{cases} t_{V_1}, & \text{if } V_1(\text{TIME}) > 0 \\ t_{tr}, & \text{otherwise (see above),} \end{cases}$$

where t_{W_1} is the time for the calculation of the function W_1 and

$$t_{r_7} = V_1(\text{TIME}).\, t^* + t_{r_5},$$

where t^* is the time for the calculation of the characteristic of the token in place m_4 (for t_{r_5} see above).

The GN E^* functions in a similar way as GN E. The tokens of the NN enter E* in place l_1. These neurons which are generated at the time of the functioning of E* appear in place m_4. A control token of the GN E_2 from the above transfers from transitions Z_4, Z_6 and Z_7 and is "a mother" of the born neurons. In the other places it functions only as a control token for the modelling process.

The place capacities of GN E^* are infinite.

The other models for functioning of NNs can be represented by similar GNs. The graphical structures of these GNs will be the same or simpler than the above ones. For example, the GN which represents the NN-back propagation method [5] will not have the transition Z_3 from the first GN described above. On the other hand some ideas for the functioning of NNs (e.g. [6-9]) will force some corrections in the form of the transition condition predicates and characteristic functions in this GN. Finally, some relations between the Petri nets and some types of the NNs are discussed in [10-12].

This paper is based on [13, 14].

REFERENCES:

[1] Jun-Wei Wong, Recognition of general patterns using neural networks, Biol. Cybern., Vol. 58 (1988), No. 6, 361-372.

[2] M. Cottrell, Stability and atractivity in associative memory networks, Biol. Cybern., Vol. 58 (1988), No. 2, 129-139.

[3] A. Fukushima, A historical neural network model for associative memory. Biol. Cybern., Vol. 50, 1984, No. 2, 105-113.

[4] A. Guez, V. Protopopsescu and J. Brahnen, On the stability storage capacity and design of continuous neural networks, Trans. SMC, Vol. 18 (1988), No. 1, 80-90.

[5] D. Rumelhart and J. McClellond, Parallel distributed processing, explorations in the microstructure of cognitron, MIT Press, Cambridge, Mass., 1986.

[6] J. Hopfield, Neural networks phisical systems with emergent collective computational abilities, Proc. Nat. Acad. Sci., 1982, 2554-2558.

[7] T. Kohonen, Self-organization and associative memory, Springer-Verlag, Berlin, 1984.

[8] L. Hadjyisky, Parallel processing in frequency dependent neural network, Parallel and Distributed Processing, Elsevier Sci. Publ., 1991, 59-74.

[9] C. Lee, Intelligent control based on fuzzy logic and neural net theory. Proc, of Internat. Conf. on Fuzzy Logic and Neural Networks, Japan, 1990, 759-764.

[10] A. Negro, R. Tagliaferri and S. Tagliaferri, Some remarks on Petri nets and neuron networks. Proc. of the IASTED Int. Symp. Applied Informatics, Grindelwald, (M. Hamza, Ed.), Anaheim, Acta Press, 1987, 28-29.

[11] T. Kristian, J. Pavlasek and P. Sapaty, Modeling of neuronal-activity using modified Petri nets, Physiologia-Bohemoslovaca, Vol. 36, No. 6, 1987, 541-551.

[12] M. Zargham, Parallel processing neural networks, Proceedings of the First Annual Meeting of the nternational Neural Network Society, 1988, Boston; Neural Networks Vol. 1, No. 1, 1988, 558.

[13] L. Hadjyisky and K. Atanassov, Theorem for representation of the neuronal networks by generalized nets. AMSE Review, Vol. 12, No. 3, 1990, 47-54.

[14] L. Hadjyisky and K. Atanassov, A generalized net, representing the elements of one neuron network set. AMSE Review Vol. 14, No. 4, 1990, 55-59.

§ 2.2: GENERALIZED NETS REPRESENTING THE TRAVELLING SALESMAN PROBLEM

Krassimir T. Atanassov

The Generalized Nets (GNs) have been used as a means for modelling real processes and theoretical problems. For example the process of solution of the transportation problem was described using a GN.

Initially, we shall construct a GN (see Fig. 2.3), describing the process of solution of a variant of the Travelling Salesman (TS) problem (see e.g. [1-7]). After this, we shall define the TS problem in a more extended form.

The temporal components of this GN will be $T = 0$, $t^{o} = 0$, $t^{*} = 1$.

The GN E consists of two subnets E_1 and E_2. The second GN E_2 has only one transition Z_1 with one place l_1. One token α with an initial characteristic "a formal description of the graph of connections between points which must be visited by the TS exists in it.

The transition Z_1 is activated at every time-moment with a duration 1. The predicate of the transition condition (there is only one) is "true". The token α receives, during its transfer from l_1 to l_1, the characteristics x^{α}_{i}, which we shall describe below.

The token β enters GN E in place l_2 with an initial characteristic "a name of vertex, where the TS has been initially".

The transition condition

$$r_2 = \begin{array}{c|ccc} & l_3 & l_4 & l_5 \\ \hline l_2 & W_1 & W_2 & W_3 \\ l_3 & W_1 & W_2 & W_3 \\ l_5 & W_1 & W_2 & W_3 \end{array}$$

where

W_1 = "From the vertex of the graph, described as an initial characteristic of α, where α is the TS, more than one path goes out which is not passed by the TS",

W_2 = "From the vertex of the graph, described as an initial characteristic of α, where α is the TS, at least one path goes out which is not passed by the TS",

W_3 = "From the vertex of the graph, described as an initial characteristic of α, where α is the TS, no path goes out which is not passed by the TS".

If a token goes in the place l_5 (i. e. the predicate W_3 is valid), then it receives as the next characteristic: "<"name of a new vertex, where the TS is going", "length of the path between the old and the present vertices", "common length of the TS transfer">".

"The length" can be changed with other coefficient.

If the predicate W_1 is also valid, each token from l_2 or l_5 splits into two tokens. The first enters the place l_5 and the second, the place l_3 without a characteristic.

Thus all tokens which determine the different directions of the TS-transfer are collected in the place l_5.

When some of these tokens do not satisfy the predicates W_1 and

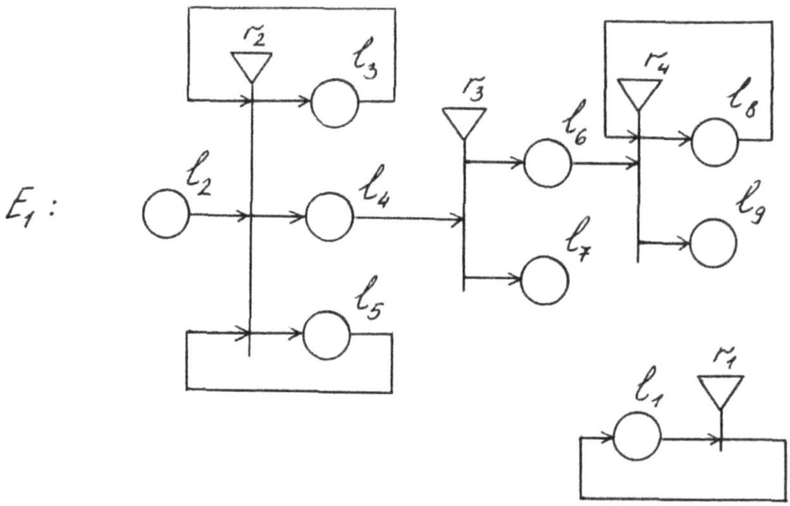

Fig. 2.3

W_3, the token enters the place l_4 without a characteristic.

The condition

$$r_3 = \begin{array}{c|cc} & l_6 & l_7 \\ \hline l_4 & W_4 & W_5 \end{array}$$

where

W_4 = "There are no untravelled vertices in the graph described as an initial characteristic of the token α "

W_5 = "There are no untravelled vertices in the graph described as an initial characteristic of the token α, but there is no path to it without repetition of the route".

Each one of the tokens which enters place l_6 receives the common length of the path as the current characteristic. The tokens, which enter place l_7 leave the net with a final characteristic "the token does not pass all graph vertices".

The transition condition

$$r_4 = \begin{array}{c|cc} & l_8 & l_9 \\ \hline l_6 & W_6 & W_7 \\ l_8 & W_8 & W_9 \end{array},$$

where

$$W_6 = (\text{"} \sum_{i=3}^{6} \bar{c}(l_i, \text{TIME}) > 1\text{" \& "} \bar{c}(l_8, \text{TIME}) = 0\text{"}) \vee \text{"}x^{\alpha}_{\text{TIME}} > x_{last}\text{"};$$

$$W_7 = \text{"} \sum_{i=3}^{6} \bar{c}(l_i, \text{TIME}) + \bar{c}(l_8, \text{TIME}) = 1\text{"} \vee \text{"}x^{\alpha}_{\text{TIME}} > x_{last}\text{"};$$

$$W_8 = \text{"}x^{\alpha}_{\text{TIME}} \le x_{last}\text{"};$$

$$W_9 = \text{"} \sum_{i=3}^{6} \bar{c}(l_i, \text{TIME}) = 0\text{"} \vee \text{"}x^{\alpha}_{\text{TIME}} > x_{last}\text{"},$$

where x_{last} is the last characteristic of the corresponding token from l_6 and x^{α}_{TIME} is the current characteristic of the token α and $\bar{c}(l, t)$ is the number of tokens in the place l at the time-moment t. The characteristic x^{α}_{TIME} is determined by the value of the current characteristic of the token which is in the place l_8. In some time-moments x^{α}_{TIME} can be "0".

The classical TS problem can be described by sush a GN. This problem can be generalized in some directions (see e.g. [1-7]). Below we shall introduce its new generalization and construct a GN representing one of them.

Let a time-scale be fixed and let T_i and T_f be the initial and final time-moments of the process, which we shall model below. It is possible that $T_f = \infty$.

Let $T = \{T_1, T_2, \ldots, T_t\}$ be a set of travelling salesmen (TSs; which is different from TS as an abriviation of "a travelling salesman").

Let $C = \{C_1, C_2, \ldots, C_c\}$ be the set of all products which the TSs delivered. Let $V = \{V_1, V_2, \ldots, V_v\}$ be the set of built-up areas (the graph-vertices) which can be visited by the TSs.

Let us use everywhere the following indices:

- $i \in \{1, 2, \ldots, t\}$,
- $j \in \{1, 2, \ldots, s\}$,
- $k \in \{1, 2, \ldots, c\}$,
- $TIME \in [T_i, T_f]$.

Below we shall define some functions related to the elements of the above sets which will determine the choices of the TSs' routes:

- $\psi(T_i, TIME) = \{C_{i,1}, C_{i,2}, \ldots, C_{i,c_i}\}$ being the set of the different types of products, which the t-th TS delivered at the current moment TIME, where

$$\bigcup_{i=1}^{t} \psi(T_i, TIME) \subset C$$

(We assume the possibility of absence of some products from the TSs' baggage at a given moment.);

- $Q(T_i, C_j, TIME)$ is the quantity of the j-th product which has the i-th TS at the moment TIME;

- $P(T_i, C_j, TIME)$ is the price of the j-th product which the i-th TS has at the moment TIME;
- $L(T_i, C_j, TIME)$ is the limit of fitness of the j-th product which the i-th TS has at the moment TIME;
- $M(V_k, TIME) = \langle K_{k,1}, K_{k,2}, \ldots, K_{k,s} \rangle$, where $K_{k,j} \geq 0$ is the quantity of the different necessary products in the built-up area V_k at moment TIME;
- $D(V_k, V_{k'})$ is the time for transfer from vertex V_k to its neighbour vertex $V_{k'}$, the equality

$$d(V_k, V_{k'}) = d(V_{k'}, V_k)$$

is not necessary;
- $E(V_k, V_{k'}, TIME)$ is the price of the transfer from vertex V_k to its neighbour vertex $V_{k'}$ at the moment TIME.
- $G(V_k, V_{k'}, TIME) \in \{0, 1\}$ determines the possibility of a tranfer from a vertex V_k to its neighbour vertex $V_{k'}$ at the moment TIME. Thus, in the frames of a given time-interval, the paths can be forbidden for transfers.
- $R(T_i, V_k, TIME)$ is the value of the gain of T_i in V_k. The value of this function can be determined e.g. as follows:

$$R(T_i, V_k, TIME) = \sum_{C_j \in \varphi(T_i, TIME)} Q(T_i, C_j, TIME).P(T_i, C_k, TIME).$$
$$sg(L(T_i, C_j, TIME)),$$

when

$$K_{k,j} \geq \sum_{C_j \in \varphi(T_i, TIME)} Q(T_i, C_j, TIME).$$

Analogously, the equality can be defined in the case when

$$K_{k,j} < \sum_{C_j \in \varphi(T_i, TIME)} Q(T_i, C_j, TIME).$$

- $N(T_i, V_k, TIME)$ is the time of staying of T_i in V_k after time-moment TIME.

Based on these definitions, we shall formulate an Extension of the Travelling Salesman Problem (ETSP). We shall also formulate other problems. The ETSP will be denoted as ETSP1.

Trace out the optimal (according to a criterion MG: "maximum gain") route of the t-th TS among the independent TSs $T_1, T_2, \ldots, T_t$ which must move in the built-up areas $V_1, V_2, \ldots, V_v$, carrying some of the products $C_1, C_2, \ldots, C_c$.

<u>THEOREM 1</u>: There is a GN in the frames of which the ETSP1 is solved.

Proof: Let the GN from Fig. 2.4 be constructed. It has the following transition conditions:

r_1 =

	l_3	l_4	l_6	l_{14}
l_1	true	false	false	false
l_2	false	false	true	false
l_3	W_1	W_2	false	W_3
l_9	W_1	W_2	false	W_3
l_{11}	false	false	true	false

where

W_1 = "there exists at least one more path along which the token can move";

W_2 = "there exists a path along which the token can be moved";

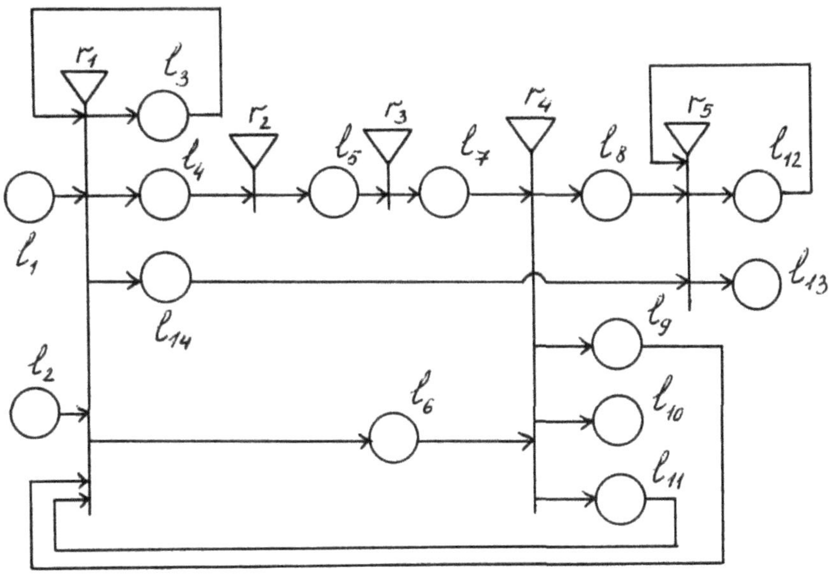

Fig. 2.4

$W_3 = \neg W_2$;

$$r_2 = \begin{array}{c|c} & l_5 \\ \hline l_4 & W_4 \end{array}$$

where

W_4 = "TIME $\geq pr_2 x_{cu}^{\alpha}$";

$$r_3 = \begin{array}{c|c} & l_7 \\ \hline l_5 & W_5 \end{array}$$

where

W_5 = "TIME $\geq x_{cu}^{\alpha}$";

	l_8	l_9	l_{10}	l_{11}
r_4 = l_7	W_6	W_7	false	false
l_6	false	false	W_6	W_7

where

W_6 = "TIME $\geq$ T + t^*",

$W_7 = \neg W_6$;

r_5 =	l_{12}	l_{13}
l_8	W_8	W_9
l_{12}	W_9	W_8
l_{14}	W_8	W_9

where

W_8 = "$\bar{c}(l_{12}$, TIME) = 0" v "$x_{cu}^{\alpha_i} \geq x_{cu}^{i}$";

where x_{cu}^{i} is the current characteristic of the α_i-token, which is located in place l_{12};

$W_9 = \neg W_8$.

Tokens α_1, α_2, ..., α_t with initial characteristics: "T_i" enter l_1 at time-moment TIME = T and after this they enter place l_3 immediately. There they receive new characteristic

"$\mu(T_i$, TIME), {<C_j, Q(T_i, C_j, TIME), P(T_i, C_j, TIME), L(T_i, C_j, TIME)> / $C_j \in \psi(T_i$, TIME)},

where $\mu(T_i$, TIME) is the vertex V_k where the TS T_i is located. Simultaneously with the above transfers, the token β enters place l_2 with an initial characteristic

$$\text{"}\{V_k,\ M(V_k,\ TIME),\ \{<D(V_k,\ V_{k'}),\ E(V_k,\ V_{k'},\ TIME),\ G(V_k,\ V_{k'},\ TIME)> / V_{k'} \in NDN(V_k)\}\text{"},$$

where $NDN(V_k)$ is the set of all next-door neighbour vertices of the vertex V_k. The α-tokens split in places l_3 (if the TS T_i has some possibilities for motion) and l_4 (if the path for motion of the TS T_i is unique). In place l_4 some tokens must be generated which have one linear predecessor. All tokens which have equal initial characteristics "T_i" are different representatives of the token α_i and there is no concurrent situation among them. Note that concurrent processes do not exist in this GN model, but such processes exist in other GN-processes of the TS-problem. The α-tokens do not receive any new characteristics in place l_3, but in place l_4 they receive the characteristic

$$\text{"}V_{k'},\ TIME + D(V_k,\ V_{k'})\text{"},$$

where $V_{k'}$ is the vertex to which the TS T_i will come. After the transfer from l_4 to l_5, token α receives the characteristic

$$\text{"}TIME + N(pr_1 x_0^{\alpha},\ pr_1 x_{cu}^{\alpha},\ TIME)\text{"}$$

and in place l_7 - the characteristic

$$\text{"}R(pr_1 x_0^{\alpha},\ pr_1 x_{cu}^{\alpha},\ TIME) - E(pr_1 x_{cu-2}^{\alpha},\ pr_1 x_{cu-1}^{\alpha},\ TIME)\text{"},$$

which is the net profit for the vertex. When the α_i-token enters place

l_9, it receives as current characteristic the name of the vertex, where it is located at the current moment, i.e. $\mu(T_i, TIME)$. In place l_8 these tokens receive as a characteristic the value of the tokens transfer in the net according to the criterion MG.

Finally, in the frames of a transition Z_5 the α_i-token with the maximum value for MG is determined.

The token β is transfered in the GN simultaneously with the transfers of the α-tokens. It enters place l_6, where it receives as a current characteristic the new values of the parameters which it has in its previous characteristic.

Thus the process of the transfer of TSs is described. For the GN's definition the following data are also necessary:

(a) the tokens transfer in a package, i.e. in the frames of one tact all tokens which can be transfered from input to output transition places, do this.

(b) the elementary time step t^o is the minimum time which is related to TS-stays or TS-transfers (in real situations, this can be e.g., 1 min.).

(c) the transitions are activated in every time-moment about the fixed time scale with parameters T_i, t^o and

$$t^*_f = T_i - T.$$

(d) the duration of the transition's activities is t^o. Therefore

$\theta_2(t) = t + t^o$.

(e) the tokens transfer is determined by their time-characteristics.

(f) the capacities of all arcs and places are ∞, without

$c(l_2) = c(l_6) = c(l_{10}) = c(l_{11}) = 1.$

(g) the priorities of all places and transitions are equal, without places l_8, l_{14} and l_{12} for which it is valid:

$\pi_L(l_8) = \pi_L(l_{14}) > \pi_L(l_{12}).$

(h) $\square_1 = \wedge(\vee(l_1, l_3, l_9), \vee(l_2, l_{11})),$

(i) $\square_4 = \wedge(l_6, l_7).$

The other transitions are of a disjunctive type.

Thus a GN in frames of which the ETSP1 is solved is constructed. This GN has some faults and the reason for them is that the number of α-tokens can grow un-limitedly.

Below we shall introduce some other modifications of the TS-problems and discuss the possibilities of their descriptions with the GNs:

(ETSP2) Trace out the locality-optimal (for a criterion LMG: "locality-maximal gain") route of the i-th TS among the independent TSs T_1, T_2, ..., T_t ($1 \le i \le t$) which must move in the built-up areas V_1, V_2 ..., V_v carrying some of the products C_1, C_2, ..., C_c when the different TSs do not know the plans of the others.

(ETSP3) Trace out the locality-optimal (for a criterion LMG): route of a in number groups of TSs: $\{T_{1,1}, T_{1,2}, ..., T_{1,t_1}\}$, $\{T_{2,1}, T_{2,2}, ..., T_{2,t_2}\}, ..., \{T_{a,1}, T_{a,2}, ..., T_{a,t_a}\}$ which must move in the built-up areas V_1, V_2, ..., V_v, carrying some of the products $C_1, C_2, \ldots, C_c$; the TSs from different sets do not concur.

The last problem can also be defined for deterministic and for non-deterministic cases.

Variants of the above problems are those where the TS must go into all the V-places or go to some of these places only; where every TS must have exactly determined qualities of products or where every TS must travel certain exactly determined distance.

The above TS problems can also be described by GNs as above.

This paper is based on [8, 9].

REFERENCES:

[1] E. Lawler, J. Lenstra, A. Rinnooy Kan and D. Shmoys (Eds.), The traveling salesman problem. A guided tour of combinatorial optimization, N.Y.: Wiley & Sons, 1985.

[2] I. Melamed, S. Sergeev and I. Sigal, The traveling salesman problem. Issues in theory. Avtomatika i telemechanika, 1989, 9, 3-33 (in Russian).

[3] I. Melamed, S. Sergeev and I. Sigal, The traveling salesman problem. Exact methods. Avtomatika i telemechanika, 1989, 10, 3-29 (in Russian).

[4] I. Melamed, S. Sergeev and I. Sigal, The traveling salesman problem. Approximate algorithms. Avtomatika i telemechanika, 1989, 11, 3-26 (in Russian).

[5] J. Little, K. Murty, D. Sweeney and C. Karel, An algorithm for the traveling salesman problem. Operations Research, Vol. 11, 1963, 972-989.

[6] T. Volgenant and R. Jonker, On some generalizations of the travelling salesman problem, Journal of Operational Research Society, Vol. 38, No. 11, 1987, 1073-1079.

[7] R. Ackoff and M. Sasieni, Fundamentals of operations research. John Wiley and Sons, Inc., New York, 1968.

[8] K. Atanassov, A generalized net, representing the travelling salesman problem. AMSE Review Vol. 14, No. 4, 1990, 61-64.

[9] K. Atanassov, Generalized net and extensions of the travelling salesman problem. AMSE Review Vol. 21, No. 2, 1992, 19-26 (in press).

§ 2.3: GENERALIZED NETS AND LABYRINTHS

Krassimir Atanassov and Rumen Christov

Different researches on the methods for motion in a labyrinth exist (see e.g., [1-3]). Here we shall describe one of them, which is based on the GNs.

Let the labyrinth L be given. It can have the form of $(m \times n)$-dimensional (0, 1)-matrix, the elements "0" and "1" of which correspond to the labyrinth corridors and walls, respectively.

We construct the GN from Fig. 2.5. It has two transitions with transition conditions

r_1 =

	l_3	l_4	l_5	m_2	m_3
m_1	false	false	false	W_1	W_2
m_2	false	false	false	W_1	W_2
l_1	W_3	W_4	W_5	false	false
l_2	W_3	W_4	W_5	false	false
l_5	W_3	W_4	W_5	false	false

where

W_1 = "$c(\bar{l}_3$, TIME) + $c(\bar{l}_5$, TIME) > 0" & "$c(\bar{l}_3$, TIME) = 0";

$W_2 = \neg W_1$;

W_4 = "end of the (unique) path or a meeting with another token on the path or along this path another token has already passed";

$W_3 = \neg W_4$;

W_5 = "there is more than one path and along this path another token has not passed";

and

r_2 =

	l_6	l_2
l_3	W_6	W_7

where

W_6 = "the token arrives to an exit";

$W_7 = \neg W_6$.

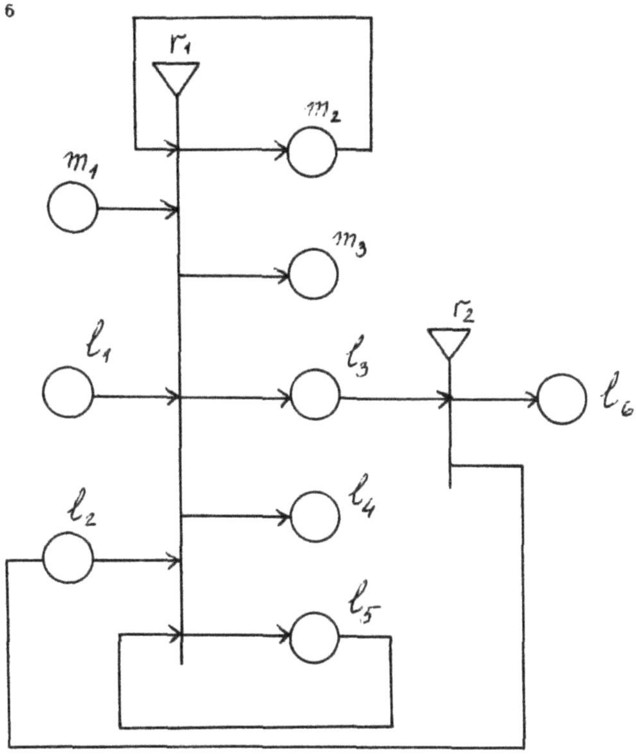

Fig. 2.5

First, tokens α and β enter the net through places l_1 and m_1, respectively and β has an initial characteristic in the form of the above mentioned matrix. If there exists exactly one path for motion, it goes to place l_3 with a characteristic "number of this path, numbered according to the hour hand". If the number of the paths is more than one, the token α splits sequentially to the tokens α_1, α_2, ..., α_s (s is the number of the different paths in front of token α). Each time one token enters place l_3 with the above characteristic another one (which will generate the next tokens) enters place l_5 without characteristic. When there are no paths for motion in front of a given token, it goes into place l_4 with the characteristic "unsuccessful end of a motion". If there is an exit in front of a token, it goes into place l_6 with the characteristic "successful end of a motion"; else the token enters place l_2 without a characteristic.

All l-places have equal priorities, which are bigger than the priorities (equal) of all m-places. Both transitions have equal priorities. All places have equal (∞) capacities, the temporal components are standard and thus not given.

The problem of motion in a labyrinth can be extended in some aspects. We shall show one extension and its GN-representation: When the labyrinth has some floors (i.e. it is not planar), in the above GN, we shall change only the characteristic function of place l_3 with "number of the path, numbered as first underneath - uphill and as second along the hour hand" and the initial characteristic of token β is $(m \times n \times k)$-dimensional $(0, 1)$-matrix.

REFERENCES:

[1] F. George, The foundations of cybernetics, Gordon and Breach Sci. Publishers, London, 1977.

[2] M. Gardner, Mathematical puzzles and diversions, Bell and Sons, London, 1965.

[3] K. Shannon, Research on theories of informatics and cybernetics, Inostrannoi Literatury, Moscow, 1963 (in Russian).

§ 2.4: FORMAL COMMUNICATION IN SCIENCE: A MODEL BASED ON GENERALIZED NETS

Radosvet Todorov and Krassimir Atanassov

§ 2.4.1 Introduction

Various studies of the communication network in science have been carried out (see [1-5]). Different models have been used to explain the informal and/or formal communication processes. Models of formal communication in science are predominantly linear and emphasize various media (articles, journals, books, etc.), participants (individuals and institutions) or functions (activities). In [1], Atherton offers two views of the communication network of scientific and technical information, one of which stresses for example the place of the publisher. Garvey presents (see [2]) a schematic overview of the scientific communication system which includes all events that occur from the time the scientist begins his research, publishes it (in a journal) until its appearance in reviews, treatises, etc. Also Garvey describes in [3] features of the current dissemination system in science and suggests some innovations in it.

Some linear models of the formal communication process are presented in a recent review article by Hills ([4]). The review looks at "the scholarly communication process in terms of the interaction and concerns of the scholar, the learned society, the publisher, the product, the librarian, and the new technology".

King et al. offer (see [5]) a schematic overview of the dissemination of scientific and technical research results by different means and of the general flow of information among scientists and engineers. The authors also supply a stochastic model of the transfer of article manuscripts between authors and publishers in order to describe the system of journal publishing in terms of cost and flow of materials.

Here we shall present at outline of a formal communication process by stressing some specific functions and participants required for the traditional transfer of article manuscripts from the author (originator) to the reader (consumer). The purpose of the study is to show a possible description of this process by means of a mathematical model with GNs.

§ 2.4.2 Schematic overview of formal communication in science

Here we choose a simplified part of the formal communication network in order to apply in a comprehensible manner the GN (having in mind the unacquainted reader). The schematic overview of the process under consideration, namely the publishing of article manuscripts and distribution of journal articles, preprints or reprints, is presented in Fig. 2.6. The author (A) submits an article manuscript to the editor(s) of a given journal. The flow of manuscripts between authors and editors is represented by arrow 1 (see Fig. 2.6). The editor (E) may reject a paper whose subject matter is inappropriate or which is "an obvious non-starter" [6]. A recommendation may be made to submit the manuscript elsewhere. This is symbolized by arrow 2. Traditionally, the editor then transmits the remaining manusrcipts "to people whom they believe to be experts in the particular topics involved" [6], this activity is represented by arrow 3. The expert or the referee (R) returns the evaluated paper (arrow 4) to the editor with his recommendations for acceptance (without change), corrections (that do not need re-examination), improvements, modifications or rejection. Only in the first case (direct acceptance) is the manuscript transferred to the publisher (arrow 5), who in our diagram is to be related with recording and reproducing activities rather than being corporated in the publishing process. If the manuscript is rejected or a revision is recommended, then the author is informed (arrow 2). The author then decides whether to revise the manuscript, to send it to another journal (usually without modifications) or to give up publication of (withdraw) his paper.

After manuscript recording, the publisher sends proof-copies to the author (arrow 6). The corrected proofs are returned for reproduction and issue preparation (arrow 7). The published articles reach the user (U) through the distributor (D), who incorporates the functions of such participants in the formal communication process as publisher, librarian, information services, etc.

Fig. 2.6 also shows the direct contact between the author and the user. The significance of arrow 9 is a reprint request and that of arrow 10 correspondingly the satisfaction of the request.

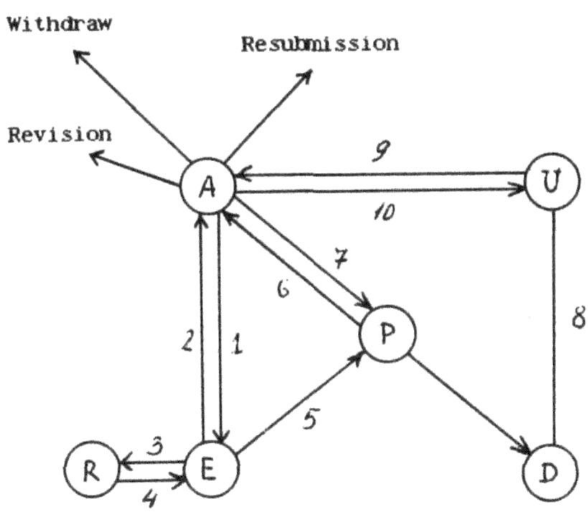

Fig. 2.6

Schematic overview of the formal communication by means of journal articles (A: Author, D: Distributor, E: Editor, P: Publisher, R: Referee, U: User)

§ 2.4.3 A GN model

Fig. 2.7 presents the graphical structure of a GN which models the relations discussed above (see also Fig. 2.6). The circled capital letters are the places where different activities are performed (editing, reviewing, etc.). f_i are "fictitious" places which are introduced to conform to the graphical representation of the GN. They are used to determine the length of time needed by the tokens to pass from a "real" to a "real" place. The f_i are "fictitious" because the tokens do not receive new characteristics as they transit through the f_i.

The token (article manuscripts) enter the net through place A (author) with the following initial characteristic "name, address of the author, manuscript's title". Place A is the entry point for all tokens which represent the manuscripts ready for a submission to a journal.

Transition condition r_1 has the trivial form

r_1 =

	f_8
f_7	true
R	true
A	true
f_4	true

.

Transition condition r_2 has the form

r_2 =

	E	f_2	f_3	W	f_4
f_8	W_1	W_2	W_3	W_4	W_5
R	true	false	false	false	false

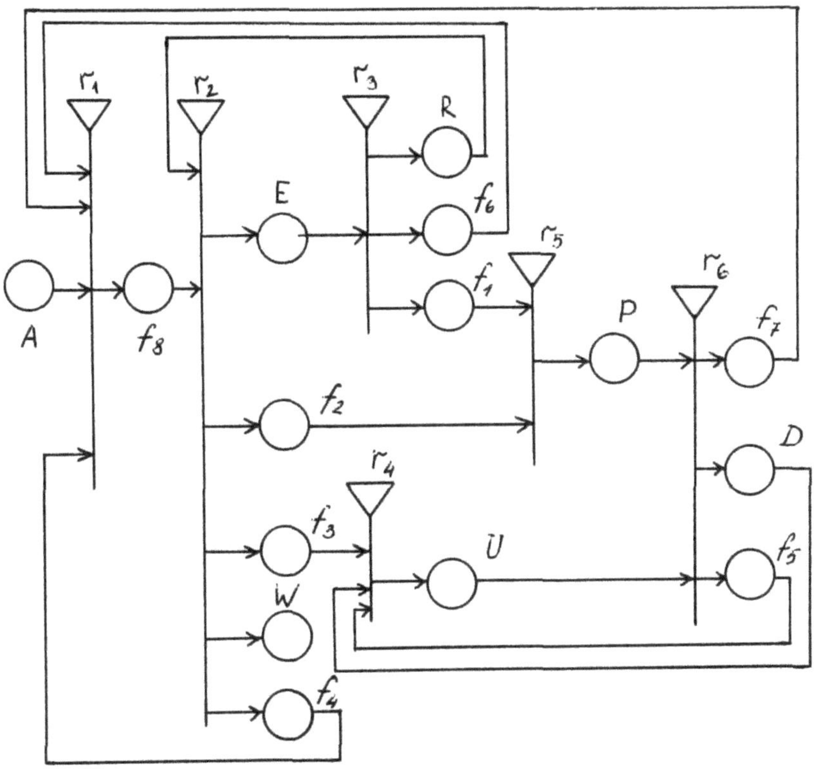

Fig. 2.7

where

W_1 = "A takes the decision to submit a manuscript";

W_2 = "A is ready to return the proof-copies to P";

W_3 = "A is willing to send a reprint to U";

W_4 = "A decides not to publish or to withdraw his manuscript"

(The place W is an output place of the model. This implies that A will take no further steps to publish the paper under consideration);

W_5 = "A agrees to make changes in order to publish the manuscript in the same journal or decides to resubmit it without improvements to another journal".

When the GN's token (which models some manuscripts) arrives in place E, it receives the characteristic "journal title".

Transition condition r_3 has the form

		R	f_6	f_1
r_3 =	E	W_6	W_7	W_8

where

W_6 = "The editor E takes the decision to inform the author A"

(The notice could include a rejection of the submitted manuscript, a recommendation for revision, etc. In other words, the token arriving in place A (coming from E) receives the characteristic "information from E to A");

W_7 = "E decides to send the manuscript to be assessed by an expert R";

W_8 = "E takes the decision to transfer the manuscript to the publisher P in order to be published (after peer reviewing)".

Transition conditions r_4 and r_5 have the trivial forms

		U
r_1 =	f_3	true
	D	true
	f_5	true

and

		P
r_5 =	f_1	true
	f_2	true

Transition condition r_6 has the form

		f_7	D	f_5
r_6 =	P	W_9	W_{10}	W_{11}
	U	W_{12}	W_{13}	W_{14}

where

W_9 = "P is ready to send manuscript copies to A" (These are the proof-copies or the reprints according to the model time),

W_{10} = "P is ready to send the issue containing the article in question to D",

W_{11} = "P is ready to send the journal issue directly to U",

W_{12} = "U wants to send a letter (with a comment) or a request for a reprint to A",

W_{13} = "U is willing to receive information or the original from D",

W_{14} = "U is willing to be disseminator of the article for another user".

Entering f_7 from P, the token has the characteristics "proof-copies" or "reprints". Arriving in D from P, it receives the characteristic "bibliographic data". When the token enters f_8 from U (through f_7) it has the characteristic "kind of information or requested material".

Let the GN function between the time-moments T and T + t^*. Initially (at the moment T) the tokens enter the net in place A and leave for E. The predicate W_1 has the current value "true" and the remaining predicates in the transition condition r_2 have zero-values. From place E the tokens move to places A or R according to the editor's judgement. Visiting places A and E the tokens receive the above mentioned characteristics.

When a token returns to f_8, it could enter place f_4 (modification of the manuscript), W (withdraw) or f_3 (sending a reprint). If the token enters R coming from E it returns obligatorily to E, and according to the characteristic received in R (referee opinion) arrives at A or P. From P it moves to A with the characteristic "proof-copies" or "reprints". The token with the characteristic "proof-copy" goes from f_8 to P through f_2. With a new characteristic (bibliographic data) the token arrives in D or f_5 coming from P. The ultimate aim of the tokens is to get to U. Notice that in our case, the place f_8 symbolizes the author's activity and U that of the reader, i.e. we do not consider (for the sake of simplicity) the possible manifestation of the user as an author and correspondingly that of the author as a reader of publications.

§ 2.4.4 Discussion

The proposed diagram of formal communication in science (Fig. 2.6) does not show all possible interactions among participants in the process. In the scheme, there could be represented other contacts as well, such as a possible direct link between the author and the referee: the author could send an accompanying note to the referee and the latter could openly publish (or answer individually) his recomendations for manuscript improvement. All the simplifications as well as some artificial divisions are made here to perceive more easily the GN as a modelling tool appropriate to communication network. The model based on the GNs is sufficiently powerful to be used for the description of the whole (complex) structure and dynamic picture of knowledge generation, representation, distribution and use, or for detailed representation of a specific part of the communication process. With minimum modifications (simplifications), the model in Fig. 2.7 can be used for example, for the description of book publishing and distribution.

Fig. 2.6 can be also presented by means of a block diagram, which would reflect the interactions among participants and the development of the process under consideration. But a block diagram cannot describe the simultaneous movement even of two or more manuscripts submitted at the same time to one, two or more journals. The simultaneous transfer of two (or more) manuscripts in the network can be described by means of e.g., the Petri nets and their other modifications, but with the increase in the number of manuscripts, it would be impossible to follow the individual fate of the manuscripts at various places of the net.

The transfer of a given number of manuscripts can be represented by the stochastic model of King [5]. However, in this case one cannot follow the dynamic development of the process or "predict with certainty the time required to perform any of the article preparation functions, nor whether the articles will be rejected, accepted, or modified" [5].

Only by means of GNs it is possible to follow the history of a manuscript (on the basis of the different GN's characteristics received by the tokens at each GN's place) and to trace its propagation in a time-frameworks.

§ 2.4.5 Possible applications of the GNs

The simulation of a formal communication process will be based on the values of the transition condition predicates, which can be determined by the token's characteristics or on the basis of experts' assessment. Using the results of the program realization and simulation, a wide variety of questions can be answered and various conclusions can be drawn. The quantitative results may include: published articles (on a certain topic or in a given journal) as a percentage of all the submitted manuscripts; or directly accepted manuscripts (i.e. without revision) as a percentage of all the accepted article manuscripts; the number of resubmitted manuscripts that are accepted; the number of submitted manuscripts by authors with the same address, etc. The qualitative answers are to questions such as the following: how does the number of submitted manuscripts on a specific topic change; is there any relation between the rejection rate and initial characteristics of the manuscript; is the flow of submitted manuscripts sufficient for the "safe" existence of a given journal, etc. Through this model many time-related questions can also be answered. One could appreciate the mean model time needed for the transfer of the manuscripts from place E to place D (publication time-lag) or the time required to pass a particular place, etc.

This paper is based on [7].

REFERENCES:

[1] P. Atherton, Views of the communication network of scientific and technical information. International Forum on Information and Documentation, Vol.1 (1975), No. 1, 10.

[2] W. Garvey et al., Research studies in patterns of scientific communication, Inf. Stor. Retr. Vol. 8 (1972), 111-122, 159-169, 207-221, 265-276.

[3] W. Garvey, S. Gottfredson, hanging the system: Innovations in the interactive social system of scientific communication. Information Processing and Management, Vol. 12 (1976), No. 3, 165-176.

[4] P. Hills, The scholarly communication process. Annual Revies for

Information Science and Technology, Vol. 18 (1983), 99.

[5] D. King, D. McDonald, N. Roderer, The journal in scientific communication: The roles of authors, publishers, libraries and readers in a vital systen. National Science Foundation Contract NSF-C-DS175-06942. Rockville, Marylands, King Research, Inc., May 1979.

[6] A. Meadows, The problem of refereeing. The Scientific Journal (A. Meadows, Ed.), Aslib Reader Series, Vol. 12 (1979), London, ASLIB, 104-111.

[7] Todorov R., Atanassov K. Formal communication in Science: A model based on Generalized nets, Scientometrics, Vol. 9 (1986), No. 3-4, 177-185.

§ 2.5: GENERALIZED NETS AND EXPERT SYSTEMS. V

Krassimir Atanassov, Peter Georgiev and Martin Tetev

The main problem in designing, implementing and using Expert Systems (ESs) is to select the proper knowledge representation formalism (see e.g. [1,2]). Most of the ESs apply the rule-based approach, i.e. they represent the knowledge in "IF situation THEN action" manner. The papers [3-6] deal with another means of representation, mainly with a generalization of Petri nets and discuss how the rule-based approach can be viewed through the new formalism with all the ensuring consequences. On the other hand the GNs are an extension of the well known Petri nets allowing a detailed description of parallel (or concurrent) processes.

The GN described in the paper can be considered as an extension of the nets from [3-6]. All notations of [3-6] are saved here. This gives the possibility to trace the development of the idea for the GN-modelling of the ESs.

On the other hand, we introduce new idea related to the ESs, based on the GN-description possibilities.

The main components of a production system are: the Data Base (DB) containing the facts about the problem which are to be solved, the Knowledge Base (KB) containing the rules which are to be used in the reasoning process, and the inference engine which operates through the KB using the DB for proving or rejecting a given hypothesis. The rules in the KB are supposed to be just in a positive conjunctive form, since otherwise the rule can be split into a number of "subrules" of that type.

We shall describe the ES's facts with new components - priorities. Now we can compare two arbitrary facts on the basis of their priorities. This possibility is very useful in the particular cases when both facts coincide or when both facts have contrary senses.

Let to every fact A of DB be juxtaposed by a natural number μ_A which corresponds to the priority of A. Let the new fact B with a priority μ_B be generated by some way in a certain time-moment when the ES functions. If both facts are not related, then the new fact enters the DB. In the ordinary ESs, the new fact B substitutes the old fact A, when B coincides with, or contradicts to A. Now the ES will function in another way, basing on the new component. When the facts A and B coincide, their representative (in particularly - A or B) stay in the DB, but with a new priority - the maximum of μ_A and μ_B. On the other hand, the fact with the maximum priority between μ_A and μ_B stays in the DB when the facts A and B are in a contradiction.

The constructed here GN (see Fig. 2.8) represents the ESs with data's priorities in the above sense.

There have been several attempts to describe production systems by some kinds of nets (see cf. [7-15]). Following [3-6] some definitions will be given here in order to describe different components of the GN appearing below.

Let a token a enters place l_1 of the GN with initial characteristic x_0^a (= "a hypothesis") and let x_{last}^a denotes its current characteristic.

Let a token b enters place l_2 with initial characteristic x_0^b (= "DB").

Let the token c enters the place l_3 with initial characterisric x_0^c (= "R"), where $R = \{R_1, R_2, \ldots, R_n\}$ is the list of the rules. Each rule R_i has the form $<C_i; A_{i,1}, \ldots, A_{i,s_i}>$, where C_i is the consequent and $A_{i,1}, \ldots, A_{i,s_i}$ are the elements of the conjunction presenting the antecedent, x_i denotes the i-th characteristic of the token with the highest priority in a given place.

Finally, let tokens $d_1, d_2, \ldots, d_s$ $(s \geq 0)$ enter place l_{40} wuth initial characteristics $x_0^{d_{cu}}$ (= <"new fact", "its priority">), where "cu" symbolises the current number of the d-type token.

The transitions in the GN are the following.

$$Z_9 = \langle \{l_{40}, l_2, l_{33}, l_{48}\}, \{l_{41}, l_{42}, \ldots, l_{46}\}, *, *, r_9, *, \square_9 \rangle$$

where

$$r_9 = \begin{array}{c|cccccc} & l_{41} & l_{42} & l_{43} & l_{44} & l_{45} & l_{46} \\ \hline l_{40} & W_{30} & W_{31} & W_{32} & W_{33} & W_{34} & false \\ l_2 & false & false & false & false & false & true \\ l_{33} & false & false & false & false & false & true \\ l_{48} & false & false & false & false & false & true \end{array}$$

where

W_{30} = "$pr_1 x_0^{d_{cu}} \in x_{last}^{\beta}$" & "$pr_2 x_0^{d_{cu}} \geq Q(pr_1 x_0^{d_{cu}})$",

where Q(A) is the priority of fact A, if $A \in x_{lasr}^{\beta}$ and Q(A) = 0, if $A \notin x_{last}^{\beta}$;

W_{31} = "$pr_1 x_0^{d_{cu}} \in x_{last}^{\beta}$" & "$pr_2 x_0^{d_{cu}} < Q(pr_1 x_0^{d_{cu}})$";

W_{32} = "$\neg pr_1 x_0^{d_{cu}} \in x^{\beta}_{last}$" & "$pr_2 x_0^{d_{cu}} \geq Q(\neg pr_1 x_0^{d_{cu}})$",

where ¬A is the negation of the fact A;

W_{33} = "$\neg pr_1 x_0^{d_{cu}} \in x^{\beta}_{last}$" & "$pr_2 x_0^{d_{cu}} < Q(\neg pr_1 x_0^{d_{cu}})$",

W_{34} = "$pr_1 x_0^{d_{cu}} \notin x^{\beta}_{last}$" & "$\neg pr_2 x_0^{d_{cu}} \notin x^{\beta}_{last}$)"

and $\square_9 = \vee(l_2, l_{33}, \wedge(l_{40}, \vee(l_2, l_{48}, l_{33})))$;

$Z_{10} = \langle \{l_{46}\}, \{l_{47}, l_{48}\}, *, *, r_{10}, *, \vee(l_{46}) \rangle$

where

r_{10} =

	l_{47}	l_{48}
l_{46}	W_{35}	W_{36}

where

W_{35} = "$c(\bar{l}_{40}, TIME) = 0$";

W_{36} = "$c(\bar{l}_{40}, TIME) > 0$"

where $c(\bar{l}, TIME)$ is the number of the tokens in place l at the current time-moment TIME;

$Z_1 = \langle \{l_1, l_{47}, l_3, l_{35}\}, \{l_4, l_5, \ldots, l_9\}, *, *, r_1, *, \square_1 \rangle$

where (here we shall use the notation from [3-6] with the corresponding corrections)

r_1 =

	l_4	l_5	l_6	l_7	l_8	l_9
l_1	W_1	W_2	false	false	false	false
l_{47}	false	false	$W_1 \vee \bar{W}$	$W_2 \vee \bar{W}$	false	false
l_3	false	false	false	false	$W_1 \vee \bar{W}$	$W_2 \vee \bar{W}$
l_{35}	false	false	false	false	$W_1 \vee \bar{W}$	$W_2 \vee \bar{W}$

where the predicates $\bar{W}$, W_1 and W_2 have the form

$\bar{W}$ = "there are more tokens before place l_1";

$W_2 = x_0^a \in x_0^b$;

$W_1 = \neg W_2$;

and $\square_1 = \wedge(l_1, l_{47}, \vee(l_3, l_{35}))$;

$Z_2 = \langle \{l_4, l_6, l_8\}, \{l_{10}^*, l_{12}, l_{13}, l_{14}, l_{15}, l_{16}\}, *, *, r_2, *, \square_2 \rangle$

where

r_2 =

	l_{10}^*	l_{12}	l_{13}	l_{14}	l_{15}	l_{16}
l_4	W_5	W_6	false	false	false	false
l_6	false	false	$W_5 \vee \bar{W}$	$W_6 \vee \bar{W}$	false	false
l_8	false	false	false	false	$W_5 \vee \bar{W}$	$W_6 \vee \bar{W}$

where the predicates W_5 and W_6 have the form

W_5 = "numb(pr$_1$ x_0^c, x_{last}^a) $\neq 0$";

W_6 = "numb(pr$_1$ x_0^c, x_{last}^a) $= 0$"

where numb(Y, y) is the number of appearances of the element y in the the ordered set Y. (In [3-6] there are other predicates related with places l_4, l_{10} and l_{12}. Here place l^*_{10} is an input to the subnet of the GN from [6]. This subnet realizes the process of checking of the hypothesis and it is aside of the present model);

$Z_6 = \langle \{l_{13}, l_{31}\}, \{l_{29}, l_{31}\}, *, *, r_6, *, \vee(l_{13})\rangle$

where

$$r_6 = \begin{array}{c|cc} & l_{29} & l_{31} \\ \hline l_{13} & W_{16} & W_{17} \\ l_{31} & W_{16} & W_{17} \end{array}$$

where the predicates W_{16} and W_{17} have the form

W_{16} = "there are no tokens in the subnet"

(the form of this predicate in [6] is similar to the above);

W_{17} = "a new fact is generated in the subnet" & "there are some tokens in the subnet";

$Z_7 = \langle \{l_{29}, l_{15}\}, \{l_{32}, l_{33}, l_{34}, l_{35}\}, *, *, r_7, *, \Box_7\rangle$

where

$$r_7 = \begin{array}{c|cccc} & l_{32} & l_{33} & l_{34} & l_{35} \\ \hline l_{29} & W_{18} \vee \bar{W} & W_{19} \vee \bar{W} & false & false \\ l_{30} & false & false & W_{18} \vee \bar{W} & W_{19} \vee \bar{W} \end{array}$$

where the predicates W_{18} and W_{19} have the form

W_{18} = "the GN-modelled ES is not a self-addapted system";

$W_{19} = \neg W_{18}$.

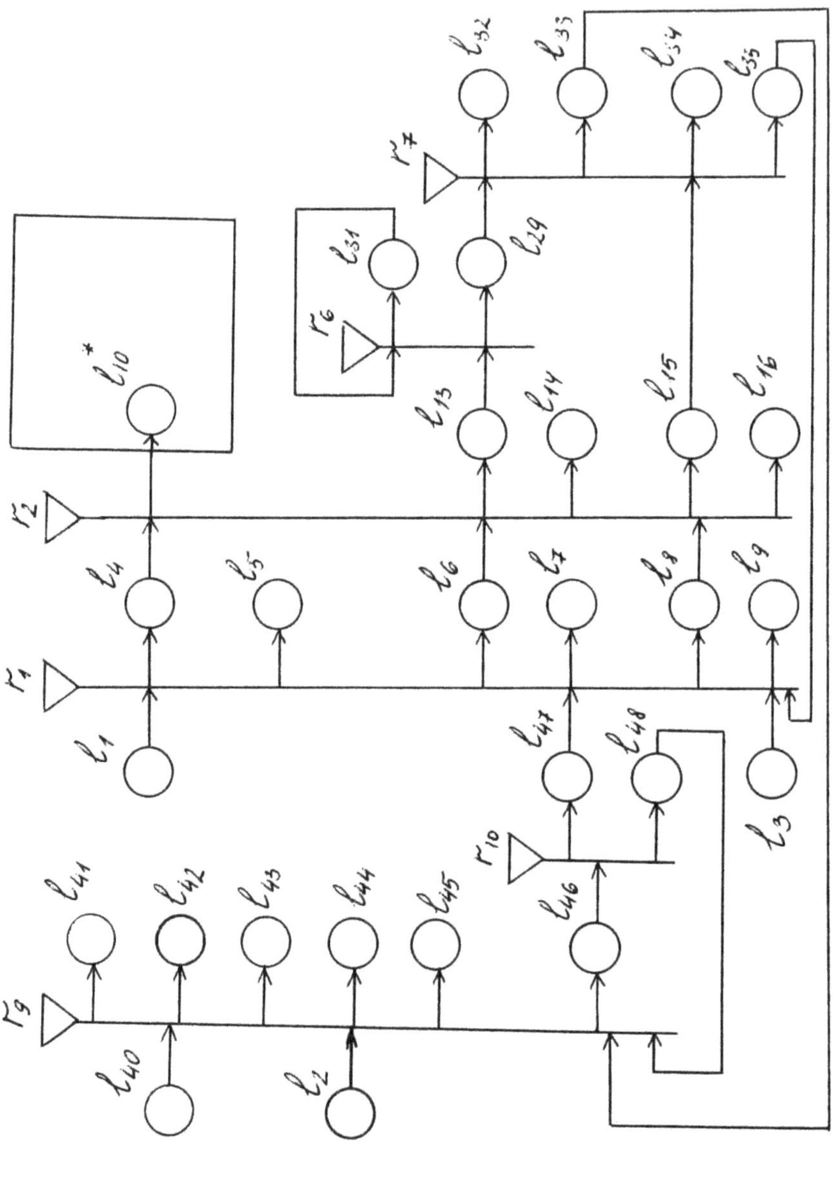

Fig. 2.8

The characteristic functions f_i of the tokens associated to the corresponding place l_i are as follows:

Φ_5 --> "$!x_0^a$" (the hypothesis x_0^a is valid);

Φ_{12} --> "$\neg(!x_0^a)$" (the hypothesis x_0^a is not valid);

Φ_{31} --> "$x_{last}^{\beta} \cup \{x\}$, where x is the last characteristic of the token which has entered place l_{20}"

(The place l_{20} is an element of the subnet. In it enter tokens with current characteristic "new fact", which is received in the process of checking of the given hypothesis x_0^{α});

$$\Phi_{46} \text{ --> } \begin{cases} (x_{last}^{\beta} - \{\langle pr_1 x_0^{d_{cu}}, Q(pr_1 x_0^{d_{cu}})\rangle\} \cup \{x_0^{d_{cu}}\}, & \text{if the token } d_{cu} \text{ is in place } l_{41} \\ *, & \text{if the token } d_{cu} \text{ is in places } l_{42} \text{ or } l_{44} \\ (x_{last}^{\beta} - \{\langle pr_1 x_0^{d_{cu}}, Q(\neg pr_1 x_0^{d_{cu}})\rangle\} \cup \{x_0^{d_{cu}}\}, & \text{if the token } d_{cu} \text{ is in place } l_{43} \\ x_{last}^{\beta} \cup \{x_0^{cu}\}, & \text{if the token } d_{cu} \text{ is in place } l_{45} \end{cases}$$

(the other functions are not defined).

The GN considered in [3-6] shows the possibility of representing production systems (their functioning and their results) by GNs. This possibility follows also from the fact that a GN can describe the functioning of Predicate/Transition nets, and in [7-15] production systems were represented by means of Petri nets, Predicate/Transition nets and other Petri net modifications. The GNs described here and in [3-6] have the following universal property: it is not dependent on the particular modelled production system.

These GNs are not connected with the description of the particular ES.

It should be mentioned that existing (known from the literature) nets representing production systems are rigidly associated with the particular system they describe and cannot be changed (deformed). On the other hand the two GNs from [12], the second GN from [3, 4] and the GNs from [5, 6] are maximally independent of the particular ES they represent. Finally, the described here GN includes as particular cases all previous constructed GNs describing ESs.

REFERENCES:

[1] Building expert systems, F. Hayes, D. Waterman and D. Lenat (Eds.) Addison-Wesley Publ. Co., Reading, Mass., 1983.

[2] D. Waterman, A guide to expert systems, Addison-Wesley Publ. Co., Reading, Mass., 1986.

[3] K. Atanassov, L. Atanassova, E. Dimitrov, G. Gargov, I. Kazalarski, M. Marinov, and S. Petkov, Generalized nets and expert systems. I. Methods of Operations Research, Vol. 59. Proc. of the 12-th Symposium on Operations Research of the Gesellschaft fur Mathematik, Okonomie und Operations Research, 1987, Passau, Frankfurt a.M: Athenaeum, 1989, 301-310.

[4] K. Atanassov, L. Atanassova, E. Dimitrov, G. Gargov, I. Kazalarski, M. Marinov, and S. Petkov, Generalized nets and expert systems. II. IFIP Symp. "Network Information Processing Systems", Sofia, May 1988, Vol. 2, 54-67.

[5] K. Atanassov, L. Atanassova, E. Dimitrov, G. Gargov, I. Kazalarski, M. Marinov, S. Petkov and M. Stefanova-Pavlova, Generalized nets and expert systems. III. Methods of Operations Research,

Vol. 63. Proc. of the 14-th Symposium on Operations Research, Ulm, Sept. 1989, 417-423.

[6] K. Atanassov, L. Atanassova, E. Dimitrov, G. Gargov, I. Kazalarski, M. Marinov, and S. Petkov, Generalized nets and expert systems. IV. Proc. of the XIX Spring Conf. of the Union of Bulg. Math., Sunny Beach, April 1990, 155-161.

[7] M. Bonacina, Petri nets for knowledge representation, Petri Nets Newsletter 27, 1987, 28-36.

[8] J. Duggan and J. Browne, ESPNET: Expert-system-based simulator of Petri nets, IEE Proc. D (Control Theory and Applications), Vol. 135, No. 4, 1988, 239-247.

[9] A. Giordana and L. Saitta, Modelling production rules by means of predicate/transition networks, Inf. Sciences 35, 1985, 1-41.

[10] U. Mainz, Netztheoretische reprasentation pradikatenlogischer begriffe und methoden, Diplomarbeit, Univ. Bonn, 1985.

[11] A. Sinachopoulos, Derivation of a contradiction by resolution using Petri nets, Petri Nets Newsletter 26, 1987, 16-29.

[12] S. Stoeva and K. Atanassov, Generalized net representation of production systems interpreters, Proc. of the Fifteenth Spring Conf. of the Union of Bulg. Math., Sunny Beach, 1986, 456-464.

[13] R. Valette, Nets in production systems, Proc. of an Advanced Course "Petri nets: Applications and Relationships to Other Models of Concurrency", Bad Honnef, in Lecture Notes in Computer Science, 255, 1986, 191-217.

[14] K. Voss, Nets in Data Bases, Proc. of an Advanced Course "Petri nets: Applications and Relationships to Other Models of Concurrency", Bad Honnef, in Lecture Notes in Computer Science, 255, 1986, 97-134.

[15] D. Koutanis, R. Rasshidi, Petri net representation of rule based expert systems. First Annual ESD/SMI Expert Systems Conference, 1987, 143-152.

§ 2.6: GENERALIZED NETS RESEARCHING THE BASIC PROPERTIES OF KNOWLEDGE BASE

Ewgeni Dimitrov, Ilya Kazalarsky and Krassimir Atanassov

One of the basic elements of every Expert System (ES) is the Data Base (DB). The structure and correctness of the DB determine the properties of the corresponding ES. The knowledge in the ES is represented as Production Rules (PRs).

The possibility to check the correctness of the KB is a very important condition for the functioning of the ES.

We shall construct a reduced GN (see App. 1) which determines the existence of at least two rules in a given DB, having identical antecedents (see Fig. 2.9). The capacities of its places and arcs are

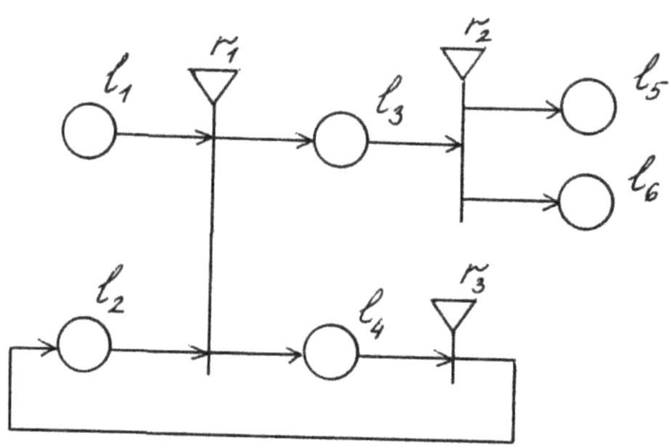

Fig. 2.9

infinite. Its transitions are as follows:

$$Z_1 = \langle \{l_1, l_2\}, \{l_3, l_4\}, r_1 \rangle$$

where

$$r_1 = \begin{array}{c|cc} & l_3 & l_4 \\ \hline l_1 & W_1 & W_2 \\ l_2 & W_1 & W_2 \end{array}$$

where

W_1 = "the place l_3 is empty" or "in place l_3 there exists at least one token with the characteristic - the same antecedent as in the present token",

$W_2 = \neg W_1$;

$$Z_2 = \langle \{l_3\}, \{l_5, l_6\}, r_2 \rangle$$

where

$$r_2 = \begin{array}{c|cc} & l_5 & l_6 \\ \hline l_3 & W_3 & W_4 \end{array}$$

where

W_3 = "the transitions Z_1 and Z_3 are not active" and "at the moment of activation of the transition Z_2 in place l_3 there exists only one token"

(this token enters place l_5 and leaves the net; if it is the unique token of the net, it receives as a final characteristic "there exist no two PRs with identical antecedents");

W_4 = "the transitions Z_1 and Z_3 are not active" and "at the moment of activation of the transition Z_2 in place l_3 there exist at least two tokens"

(these tokens enter place l_6 and leave the net with a final characteristic "the PRs which are initial characteristics of these tokens have identical antecedents");

$$Z_3 = \langle\{l_4\},\ \{l_2\},\ r_3\rangle$$

where

$$r_3 = \begin{array}{c|c} & l_2 \\ \hline l_4 & W_5 \end{array}$$

where

W_5 = "the places l_1 and l_2 are empty".

All PRs of a given KB can be described as initial characteristics of the tokens of the above GN and thus an answer for the question of the existence of identical antecedents in PRs can be obtained.

The GN which checks the set of all PRs of a given KB about the existence of at least two rules from the form $A \vdash C$ and $A \& B \vdash C$ has the form as in Fig 2.10. The transitions Z_1 and Z_3 in both nets coincide with those from Fig. 2.9 without

W_1 = "the place l_3 is empty" or "in place l_3 there exists at least one token with the characteristic - the same consequent, as in the present token".

The transition

$$Z_2 = \langle\{l_3,\ l_9\},\ \{l_5,\ l_6\},\ r_2\rangle$$

where

$$r_2 = \begin{array}{c|cc} & l_5 & l_6 \\ \hline l_3 & W_6 & W_7 \\ l_9 & W_6 & W_7 \end{array}$$

where

W_6 = "the place l_5 is empty" or "in place l_5 there exists at least one

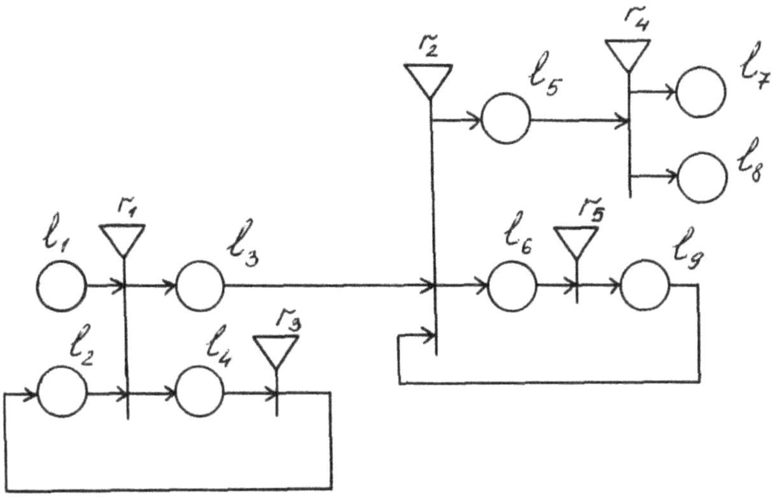

Fig. 2.10

token with the characteristic - an antecedent, being a subterm of the antecedent which is an initial characteristic of the present token",

$W_7 = \neg W_6$.

The transition

$$Z_4 = \langle \{l_5\}, \{l_7, l_8\}, r_4 \rangle$$

where

$$r_4 = \begin{array}{c|cc} & l_7 & l_8 \\ \hline l_5 & W_3 & W_4 \end{array}$$

where W_3 and W_4 are as above.

The transition

$$Z_5 = \langle \{l_6\}, \{l_9\}, r_5 \rangle,$$

where

$$r_5 = \begin{array}{c|c} & l_9 \\ \hline l_6 & W_8 \end{array}$$

where

W_8 = "places l_3 and l_5 are empty".

The result of functioning of this GN is as follows: the tokens corresponding to the rules of the type $A \vdash C$ and $A \& B \vdash C$ (if they exist) leave the net at place l_8. If all tokens leave the net from place l_7, then in the KB there exist no such rules.

The GN checking a given KB for contradictions, i.e., for simultaneous existence of rules of types $A \vdash B$ and $A \vdash \neg B$, is similar to the GN from Fig. 2.9, but here the consequent is $C = \neg B$.

The GN which checks a given KB for existence of two rules of types $A \vdash B$ and $B \vdash A$ also has the form as in Fig. 2.9 and all notations are identical without the predicate

W_1 = "the place l_3 is empty" or "in place l_3 there exists a token for which the antecedent (in its initial characteristic) is a consequent in the initial characteristic of the present token and the consequent (in its initial characteristic) is an antecedent in the initial characteristic of the present token".

As the result of functioning of this GN the tokens leaving the place l_6 have initial characteristics - contradicting rules. If all tokens leave the net at place l_5, then the KB is noncontradicting.

Considering the GN, we conclude that such nets are a useful graphic tool for representing the structure and functioning of the

ESs. The representation of the ESs by GNs has certain advantages and it will probably become even more popular in future. This will contribute to an easy transition from present systems to systems of the fifth generation computers.

This paper is based on [1].

REFERENCE:

[1] E. Dimitrov, I. Kazalarski and K. Atanassov. Generalized net representing the basic properties of knowledge base. First Scientific Session of the "Mathematical Foundation of Artificial Inteligence" Seminar, Sofia, Oct. 10, 1989, Preprint IM-MFAIS-7-89, Sofia, 1989, 12-14.

§ 2.7: PARALLEL KNOWLEDGE PROCESSING MODELLING IN EXPERT SYSTEM USING GENERALIZED NETS

Ivan Hristozov and Evgeni Tzolov

The main task in designing expert systems (ES) for application in fields such as decision support, image processing, real time control and others is to achieve high-speed data processing. The application of the parallel computation principles in the ES as a whole and in the inference mechanism in particular, is a way to achieve a quick response.

A method for a formalized description of a parallel inference mechanism in ES based on production rules using Generalized Nets (GNs) is proposed in this paper. The organization and the interference of the parallel processes in a multiprocessor system are being studied. The system is built up of specialized processors for parallel computations (SPPC) of the transputers type [1-3].

The production systems are a popular means for knowledge representation. Each production system consists of three basic components: a data base, a set of rules (a knowledge base) and a rule interpreter (a logical inference mechanism). The knowledge is represented by production rules of the type

IF "condition 1", "condition 2"..."condition N", THEN "conclusion".

The activation of each rule changes the data in the data base.

Parallel data processing in ES on the basis of a production system will be considered as a simultaneous interpretation of a set of rules in a multiprocessor system of SPPC. The rule evaluation in each SPPC is sequential.

A list of identificators (names) of the conditions and the conclusions included in the rules and a list of identificators of the hypotheses to be proved are entered in the process of the knowledge base formation. The list of hypotheses is a subset of the list of the conditions' and conclusions' identificators. After forming the knowledge base only the rules necessary for proving the given hypotheses are separated.

The knowledge base thus reduced is divided into groups (levels) of rules, which can be activated parallely, i.e. they are not informationally interconnected. Rules whose conditions (the variables on their left sides) are not conclusions of other rules, i.e. rules that can be directly evaluated, are referred to as the first processing level. Rules that depend on at least one rule of the (i-1)-th level are referred to as the i-th level ($i \geq 2$). Rules of the i-th level may depend on rules of the lower levels. Only the rules whose conclusions are the chosen hypotheses are evaluated on the last level. The rules from a given level are distributed for parallel processing between the processors of the multiprocessor net.

For the case where the number of independent rules of a given level is less than or equal to the number of processors in the net, each rule can be evaluated on a separate processor. When the number of rules on a level is greater than the number of processors, some of the independent rules cannot be evaluated parallelly, because of the lack of resources (processors).

If the net of processors is homogeneous (built up of similar processors), one rule will be evaluated at one and the same time on the different processors. So balance processor loading in time may be chosen as a distribution criterion. The distributed rules (the individual knowledge bases) are loaded in the processors.

When starting up the inference mechanism, the data base is sent to each processor of the net. When processing the rules of a given level, each processor changes a given part of it. In order to standardize the data base at the end of the rule processing on a given level, information exchange between the SPPC is necessary (the conclusion values received in each SPPC are necessary to continue the processing in the other processors of the net).

The evaluation process of the rules continues until there are no more levels. The values of the chosen hypotheses are sent to the user (the controlled object), new data (if any) are received from him, and the processing is repeated.

When developing the multiprocessor system for parallel processing, the question as to the evaluation of its effectiveness arises. On the stage of designing when the system is still not physically realized, this is possible by using the imitation modelling methods.

The GNs are introduced as an effective means for modelling and simulating. They permit a detailed investigation, a natural way of representing parallel and asynchronous processes, receiving temporary characteristics.

The operation of the multiprocessor system performing parallel inferece in the above stated way may be described by a reduced GN (see App. 1), given on Fig. 2.11. It has three transitions. Transition Z_1 is of the type

$$Z_1 = \langle \{l_1, l_2, l_3, l_4, l_7, l_8, l_{14}\}, \{l_8, l_4, l_9, l_5, l_6, l_{10}, l_{13}\}, r_1 \rangle.$$

In the beginning of the GN functioning, the place l_1 contains tokens, symbolizing the rules of the knowledge base. Each token has the following characteristics: "execution time, "priority". The priority locates the level of a rule. The rules with an identical priority are processed for one realization of the transition. The tokens from place l_1 with the lowest priority pass into place l_{13} with the characteristics - "rule number", "identificator showing whether the operation in the THEN part is performed or not". Tokens with a higher priority pass in place l_8 to without changing their characteristics and are

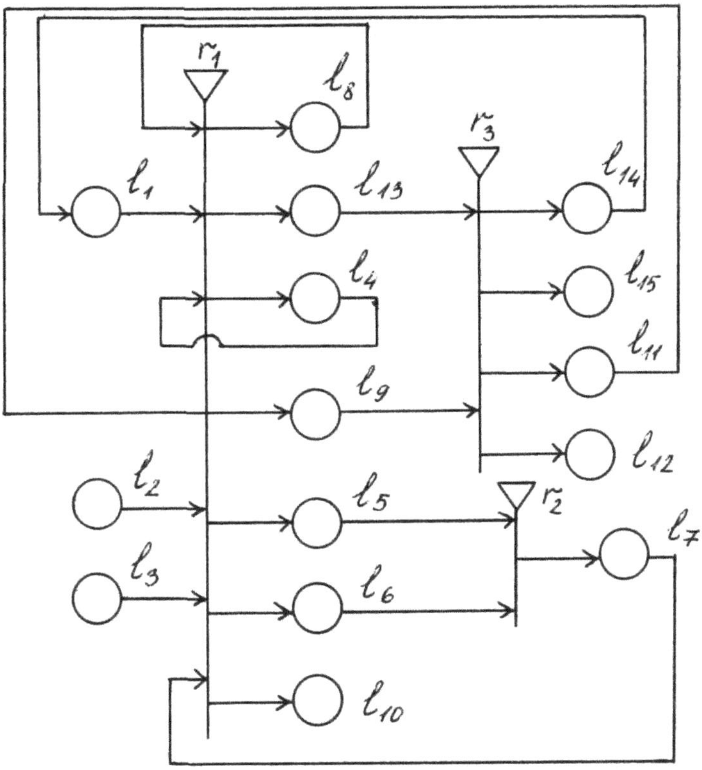

Fig. 2.11

subjected to a subsequent processing. After terminating the logical inference all rules are evaluated and the tokens from l_1 have passed into l_{13}. The tokens from l_{13} go into place l_{15}, if a history tracing of the inference is necessary, or into l_{14} for a cyclic repetition of the inference.

In the beginning of the GN functioning, place l_2 (input place) has tokens with initial characteristics "data base identificator" and a "list of hypotheses we are interested in". Each SPPC has its own data base identificator. These tokens pass into place l_4 with the characteristic "updated data base". Thus l_4 becomes the input place for the next processing level. When there are no more levels, the tokens from place l_2 and l_4 pass into place l_9 with the characteristic "final data base". In the beginning of the GN functioning place l_3 has tokens whose characteristics are: "number of the corresponding processor", "numbers of the processors directly connected with it", "identificator of the data base, which is the initial characteristic of the token from l_2".

When the processing of rules with the same priority ends, the tokens from l_3, under the conditions specified below, pass into places l_5 and l_6, with the respective characteristics: "numbers of processors from which data are received", "numbers of processors, to which data are sent", besides "time for communication between the corresponding processors". In the case where the number of rules of a given level is smaller than the number of processors in the net, some processors do not take part in the processing on that level. The tokens with characteristics "the numbers of such processors" pass into l_6 and have no characteristics. The tokens from l_3, corresponding to operating processors only, pass into place l_5 with characteristics "new facts (changes in the data base received during the operation of the corresponding processor with the rules of the specified level)". So, when passing the transition Z_1, the tokens from l_3 are divided into new tokens, the number of which is equal to the number of transmissions/receivings . In l_7 all processors have the characteristic "new facts, received during operation".

At the end of the operation the tokens from l_3 pass into l_{10} and have no characteristics.

The condition r_1 is of the type

r_1 =

	l_4	l_5	l_6	l_8	l_9	l_{10}	l_{13}
l_1	false	false	false	W_5	false	false	W_4
l_2	W_1	false	false	false	W_4	false	false
l_3	false	W_2	W_3	false	false	W_4	false
l_4	W_1	false	false	false	W_4	false	false
l_7	false	W_2	W_3	false	false	W_4	false
l_8	false	false	false	W_1	false	false	false
l_{11}	W_1	false	false	false	W_4	false	false
l_{14}	false	false	false	W_5	false	false	W_4

where

W_1 = "a new fact is found",

W_2 = "there are processors that will transfer data";

W_3 = "there are processors that will receive data";

W_4 = "all rules are evaluated";

W_5 = "there are groups of rules for processing".

The appearance of a token in place l_9 activates the transition

$$Z_3 = \langle \{l_9, l_{13}\}, \{l_{11}, l_{12}, l_{14}, l_{15}\}, r_3 \rangle,$$

which defines data exchange with an external object. The token from

l_9 is divided in two: one part passes into place l_{11} with characteristic "input data identificators" and receives data, if any, from the final user; the other part passes into place l_{12} with the characteristic "hypotheses" and is sent to the user (controlled object). The condition r_3 is of the form

$$r_3 = \begin{array}{c|cccc} & l_{11} & l_{12} & l_{14} & l_{15} \\ \hline l_9 & W_6 & true & false & false \\ l_{13} & false & false & W_7 & true \end{array}$$

where

W_6 = "there are new data for receiving";

W_7 = "a new inference execution is necessary".

The transition Z_2 is activated if there is at least one pair of processors ready for data exchange between them. It unites tokens with the same initial characteristic "processor number". By the end of the last exchange operation, all tokens from l_5 and l_6 have passed into place l_7, which is the initial place for processing the next level of rules with one and the same priorities. Transition Z_2 has the form

$$Z_2 = \langle \{l_5, l_6\}, \{l_7\}, r_2 \rangle$$

with condition

$$r_2 = \begin{array}{c|c} & l_7 \\ \hline l_5 & W_8 \\ l_6 & W_9 \end{array}$$

where

W_8 = "the processor in 16 has received all data";

W_9 = "the processor in 15 has transferred the new facts to the other processors".

If a moment of time t_1 is compared with the moment of activating the transition Z_1, and a moment t_2 to the moment of the end of exchange with the controlled object (deactivating the transition Z_3), the difference $t_2 - t_1$ determines the time for data processing in the expert system.

The model developed permits on to evaluate the characteristics of logical inference parallel computers with a different connection topology, number of processors, rule distribution, exchange organization (priority of one processor to another) aiming at finding an optimal (according to a specified criterion) version.

REFERENCES:

[1] D. Richard, C. Askew and D. Carpenter, Practical parallelism using transputer arrays, Lecture Notes in Computer Science 258, 1987.

[2] Transputer development system 2.0, IMS D700C, INMOS, 1987

[3] K. Yanev, D. Todorov, N. Avramov and E. Elitzina, Designing digital devices with microprocessors, Sofia, Tehnika, 1987 (in Bulg.).

§ 2.8: GENERALIZED NETS AND DISTRIBUTED DATABASE TECHNOLOGY

Alexander Georgiev

§ 2.8.1. Introduction

This article is an introduction to present our Generalized Net (GN) approach to the Distributed Database Technology (DDBT). The GNs should be considered as a possible response to the problem of the development of tools for parallel process modelling.

The DDBT is a state-of-the-art technology in the database theory. The DDBT consists of two basic domains: Global Intelligence (GI) with laws and rules governing the Distributed Database (DDB) [1].

The aim of this article is to introduce the foundations of the GN approach to the building of the Global Distributed Database Management System (GDDBMS) in the DDBT research area. Our approach follows the specifications of the ANSI/SPARC project proposed in [2]. Following the current status of the DDBT theory, the GDDBMS is considered to have two conceptual domains: the DDB as a set of new basic structural concepts in the DDBT as well as a lower conceptual domain, and an upper conceptual domain - the GI as a wise manager in a distributed environment. All discussions are on the conceptual level only. Finally, we propose the architecture of an abstract GDDBMS which integrates all features from the DDBT area on the common basis of the GN formalism.

*

* *

§ 2.8.2. Distributed database - basic structural concepts

In the following paper we shall discuss three basic concepts and their consequences for organization of the DDB logical structure of the information network. They define the data architecture in the DDB on some layers or schemes, as shown in [1].

The first notion is fragmentation. Fragmentation is a scheme which describes how tables (relations) are shared among the network nodes logically. There are three fragmentation types: horizontal, vertical and mixed one. Fragmentation allows one to define different users' views on the data in the database.

The second notion is replication. Replication is a scheme which describes the distribution of different whole tables as copies of data among the network nodes. Replication refers to fast local access to data as well as to a higher assurance of the network's working.

The third notion is allocation. Allocation of data is a combination of fragmentation and replication. Allocation is a scheme which describes the fragments with their physical addresses on the network. Allocation is a key element in a project of DDB. Allocation has a guiding principle: local availability.

The three concepts shown above define the layer's architecture of the DDB. The DDB architecture has to follow the basic principles for logical and physical independency of the data. Therefore, there should exist mappings between the layers. Mapping is a description of facts and rules for a transformation from one logical aspect to another logical or physical aspect of the DDB. In that sense, a mapping could be viewed as a global, fragmentated, allocated, or local-mapping scheme.

*
* *

§ 2. 8. 3. Distributed database - basic logical concepts

The notion "distributed transparency" is closely related to the DDB architecture. Each layer from the architecture is associated with a conveniently distributed transparent layer. A transparency of the DDB means that the user should see that the database resided at the user's node with its whole entity. The distributed transparency is orientated to ensure both logical and physical independency of the data in the DDB. The fragmentation transparency has the highest transparency level.

The Distributed Data Integrity (DDI) is a feature that ensures the consistency and security of the data. It is related to the following problems of supporting the activities on the DDB: query optimization, concurrency control and transaction handling.

*
* *

§ 2.8.4. Global intelligence

We consider GI as an entity which is most important for the the existence of the DDB. A node contains a Database Management System (DBMS). The DBMS represents Local Intelligence (LI) in the DDB system. The GI exists above the LI. GI could be considered as a Distributed DBMS (DDBMS). The GI manages and controls the data structure, the row data, knowledge about the data (metadata) and all activities of the overall DDB. The GI is responsible to ensure:

(a) Distribution Transparency as a structural aspect as well as an activity in the DDB;

(b) a DDI by means of supporting the query optimization, concurrency control and the transaction handling.

All these activities require the capacity to maintain parallel processing. Parallel processing is related to the notion of real-time synchronization. It is the biggest problem in a DDBMS. For solving the problem it is necessary to have a power modelling tool for behavioural research of a parallel processing in the real-time situation. The GNs are the ideal ones in this sense. The GN can behave as a set of processes at a certain moment of time, each one with its own history, local time and constraints. Moreover, the GNs can forecast the behaviour of a stand-alone process, a subset of related processes and the processes as a whole, and then detect deadlock or wrong situations as well as correct them. We consider the GNs a conceptual modelling tool for a GI above the DDB. We distinguish some sub-models, for example: GN-model for query decomposition, GN-model for query optimization, GN-constraint integrity model, GN-transaction handling model, and GN-conceptual model of the GI (GNGI-model), which is a general one.

*

* *

§ 2. 8. 5. Conclusions

In this paper we have presented the idea of the GN application in the DDBT area. We have reason to think that this is a new approach to the building of a DDBMS as a GI in a distributed environment. We will discuss the submodels of the GNGI-model precisely in the future works. Moreover, the idea of using the GN as a modelling tool in the DDB area could reflect the performance of intelligent and power database machines.

REFERENCES:

[1] R. Davis, Sharing the Wealth, BYTE, September 1989.

[2] The ANSI/SPARC DBMS framework, Information Systemsm, Vol. 3, No. 3, 1978.

Chapter 3:

SOME APPLICATIONS OF GENERALIZED NETS IN ECONOMICS, INDUSTRY AND TRANSPORT

Several GN-models describing various processes in areas of economics, industry and transport are introduced. Some of them are published for the first time (§ 3.3, § 3.4, § 3.5, § 3.7, § 3.8). These models describe the processes from the view of Bulgarian economics, industry and transport. Some of them give only an idea about the form of the GN-models which can be constructed (e.g. in § 3.3), others describe the processes in a broad outline (§ 3.1, § 3.4, § 3.5, § 3.7, § 3.8). The model from § 3.2 and those from § 3.6 are more detailed. The models from § 3.7 reflect the first steps made on the global modelling of the NEFTOCHIM Petrochemical Combine in Bourgas (Bulgaria). Therefore, the GN-models described here are not universally valid or of worldwide importance. On the other hand, they do illustrate the possibility for a GN-description of such processes and also for other similar ones. The models are realized in the frames of the current version of the program package for GNs and they can be given to those who are interested in them.

§ 3.1: MODELLING THE ACTIVITIES OF INTERNATIONAL TRANSPORT MARKET BY GENERALIZED NETS

Antoaneta Kirova and Krassimir Atanassov

The international transport market is a complicated dynamical system developing under the influence of a number of factors. Its study is connected with creating different models which give a visual aspect of the processes and afterwards make it possible to determine the ways of improving them in practise.

The modelling of economic processes includes a number of theoretical and statistical data concerning their appearance in reality. The models give an idea about the interconnection of ways of arraanging the elements of a complicated economic system. The gears and tasks of the management at a given stage define a number of ways of modelling: statical, dynamical, expert-immitational and so on. The statical modelling is characterized by firmly set out conditions which surround the economic process. An example most typical in this aspect is the transportation task. The creation of dynamical models is possible with non-stopping processes or phenomena. The widespread use of computers in practise gives the possibility of modelling those processes and phenomena which only the expert-immitational method can describe. This concerns a number of management processes, parallel-going processes, and so on. One of the ways of creating expert-immitational and simulational models is by using different types of nets.

The main quality of GNs is that they give not only a time measurement of the tokens, but also the conditions for going from one place to another. This creates some prerequisites for simulating real processes and for optimising the existing ones.

Here we intend to show two of the processes connected with the international market of transport services. The modelling concerns the activities of the forwarding agent and the transporting firm.

*
* *

Let the GN of the forwarding agent be E_c (see Fig. 3.1). We shall assume that the tokens represent the different transportation means conducted by the forwarding agents. These tokens possess the following characteristics: type of the transportation means, the (current) locations, time for covering a certain distance and price of the transportation service. All these are changing in time and in the universe when passing from one place to another.

The forwarding agent's (reduced) GN is as follows:

$$E_c = \langle\langle A, \pi_A, f\rangle, K, \langle T, t^o, t^*\rangle, \langle X, \Phi, b\rangle\rangle,$$

where

(a) $A = \{Z_1, Z_2, \ldots, Z_6\}$,

(b) the set K of tokens consists of all the trailers and other transportation means which the forwarding agent might use at a moment of time.

(c) the time GN's components are as follows:

T - a fixed start time-moment,

t^o - one day,

t^* - one year.

The token's characteristics, which are related to the time are a very important part of the generated information in the net. The initial token's characteristics have the form <"type of the transportation means", "location">. The characteristic function Φ will be illustrated at each place after its description. The GN has 12 places and for them, the function Φ juxtaposes to the corresponding tokens

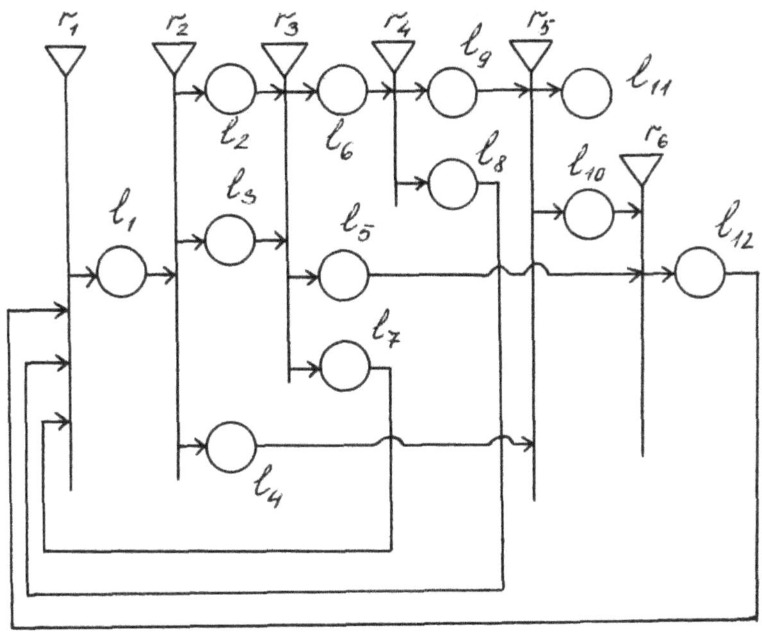

Fig. 3.1

the following values:

l_1 - "location of the transportation means",

Φ_1 - <"new (current) location", "price of passage">,

l_2 - "transportation means sent for loading",

Φ_2 - <"kind of a good", "price of passage", " time for passage">,

l_3 - "transportation means of sending to the agent's warehouse",

Φ_3 - <"kind of a good", "price of passage", " time for passage">,

l_4 - "transportations means of sendinf for re-loading",

Φ_4 - <"direction of transportation">,

l_5 - "transportation means for proceeding to the consumer",

Φ_5 - <"direction of transportation">,

l_6 - "transportation means for a mixed transportation",

Φ_6 - <"location", "time for loading operations", "price of passage">,

l_7 - "transportation means sent to an intermediate's warehouse",

Φ_7 - <"location", "time for loading operations", "price of passage">,

l_8 - "transportation means is discharged",

Φ_8 - <"direction of transportation">,

l_9 - "trailer/road railer on mixed mode of transportation",

Φ_9 - <"direction of transportation">,

l_{10} - "trailer/road railer at the final point",

Φ_{10} - <"location", "time for loading operations", "price of passage">,

l_{11} - "emergency situation",

Φ_{11} - <"emergency situation">,

l_{12} - "goods in the consumer's warehouse",

Φ_{12} - <"location", "time for loading operations", "price of passage">.

The above mentioned places and the characteristic functions related to them, correspond to the activity of the forwarding agent, because he represents the goods' owners to the transportation firms. He is concluding the transportation contracts, determines the mode of transportation and is used to differentiate other operations connected with the preparation, running and completion of the transportation process.

Components which are anticipated to have an indirect participation in the model are omitted. They are t_1-, t_2- and M-transitions' components, π_L-, c-, θ_1-, θ_2-, π_K-, θ_K-, b-GNs' components. Therefore, the described GN is one that is a reduced from the class

$$\Sigma^{A_3, A_4, A_6, \pi_L, c, \theta_1, \theta_2, \pi_K, \theta_K, b}.$$

The capacities of the places and the arcs are infinite. The transitions have the following forms:

$Z_1 = \langle\{l_7, l_8, l_{12}\}, \{l_1\}, r_1, v(l_7, l_8, l_{12})\rangle,$

where

$$r_1 = \begin{array}{c|c} & l_1 \\ \hline l_7 & true \\ l_8 & true \\ l_{12} & true \end{array}.$$

The transition's type of Z_1 is "v" because it can be fired if a token (a transition mean which is already discharged) appears in one of its input places (i.e. in l_7, l_8, l_{12}). This transition corresponds to the "sending of the transition means from its owner to the forwarding agent".

$Z_2 = \langle\{l_1\}, \{l_2, l_3, l_4\}, r_2, v(l_1)\rangle,$

where

$$r_2 = \begin{array}{c|ccc} & l_2 & l_3 & l_4 \\ \hline l_1 & W_1 & W_2 & W_3 \end{array}$$

where

W_1 = "the transportation means is with the producer",

W_2 = "the transportation means is within the agent's warehouse",

W_3 = "the transportation means is sent for loading";

$Z_3 = \langle \{l_2, l_3\}, \{l_5, l_6, l_7\}, r_3, \wedge(l_2, l_3) \rangle$,

where

$$r_3 = \begin{array}{c|ccc} & l_5 & l_6 & l_7 \\ \hline l_2 & W_4 & W_5 & W_6 \\ l_3 & W_7 & W_8 & false \end{array}$$

where

W_4 = "the transportation means is moving from the producer to the consumer",

W_5 = "the transportation means is for re-loading",

W_6 = "the transportation means is in the intermediate's warehouse",

W_7 = "the transportation means is moving from the agent's warehouse to the consumer",

W_8 = "the transportation means is for re-loading in the agent's warehouse".

Here l_2 and l_3 are input and l_5, l_6, l_7 are output places. The transition's type is "$\wedge$", because for its activation it is necessary to have at least one token in each of the transition's input places. Z_3 means "first stage of transportation".

$Z_4 = \langle \{l_6\}, \{l_8, l_9\}, r_4, \vee(l_6) \rangle$,

where

$$r_4 = \begin{array}{c|cc} & l_8 & l_9 \\ \hline l_6 & W_9 & W_{10} \end{array}$$

where

W_9 = "the transportation means is discharged",

W_{10} = "trailer/road railer continuing".

Z_4 denotes "end of the first stage of transportation and beginning of the second one.

$Z_5 = \langle \{l_9, l_4\}, \{l_{10}, l_{11}\}, r_5, \wedge(l_9, l_4) \rangle$,

where

$$r_5 = \begin{array}{c|cc} & l_{10} & l_{11} \\ \hline l_9 & W_{11} & W_{12} \\ l_4 & W_{13} & W_{14} \end{array}$$

where

W_{11} = "the goods have arrived",

$W_{12} = \neg W_{11}$,

W_{13} = "the transportation means have arrived",

$W_{14} = \neg W_{13}$.

Here l_9 and l_4 are input, and l_{10} and l_{11} are output places. The transition's type is "$\wedge$" because for its activation it is necessary to have at least one token in each of the transitions' input places. Z_5 means "end of the second stage of transportation and beginning of the third one".

$Z_6 = \langle\{l_5, l_{10}\}, \{l_{12}\}, r_6, \wedge(l_5, l_{10})\rangle,$

where

$$r_6 = \begin{array}{c|c} & l_{12} \\ \hline l_5 & true \\ l_{10} & true \end{array}.$$

Z_6 expresses "completion of the transportation process and delivery of goods to the consumer".

The proposed model shows only those parts of the forwarding agent's activity which require management in time and universe. This is particularly important in the international field, where the delivery term needs punctuality in the contacts between the goods and the transportation means. The effectiveness is greater in organizing the multimodal transportation abroad, for the processing there is divided into different stages (land - river - land, land - see - land, etc.).

*

* *

The transporter's GN E_n is as follows:

$E_n = \langle\langle A, \pi_A, f\rangle, K, \langle T, t^o, t^*\rangle, \langle X, \Phi, b\rangle\rangle,$

where

$$A = \{Z_1, Z_2, \ldots, Z_6\}$$

(same as above, but the new transitions are different from the above); the elements of the set of tokens K also represent the transportation means in the transportation firm; the time components of the net are the same as with E_c; the token's characteristics in E_n are analogical to E_c; the initial token's characteristics are from the type: <"type of the transportation means", "location">.

The places of this net denote the following:

l_1 - "the transportation means is ready for loading",

l_2 - "the transportation means is sent for loading",

l_3 - "the transportation means meets goods from a multimodal transport",

l_4 - "the transportation means has got the goods",

l_5 - "the transportation means has given the goods to another transport",

l_6 - "the transportation means is discharged",

l_7 - "the transportation means is moving loaded",

l_8 - "the transportation means has left the goods to the consumer",

l_9 - "an emergency situation",

l_{10} - "end of the second stage of a multimodal transportation",

l_{11} - "an emergency situation with a multimodal transportation",

l_{12} - "continuing of the second stage of a multimodal transportation",

l_{13} - "start of the third stage of a multimodal transportation",

l_{14} - "repair/storage of the transportation means".

The function Φ of the places is similar to the above:

Φ_1 - <"location", "time", "price of passage">,

Φ_2 - <"direction of transportation">,

Φ_3 - <"direction of transportation">,

Φ_4 - <"direction of transportation">,

Φ_5 - <"direction of transportation">,

Φ_6 - <"direction of transportation">,

Φ_7 - <"direction of transportation", "time", "price of the passage">,

Φ_8 - <"direction of transportation">,

Φ_9 - <"emergency situation">,

Φ_{10} - <"location", "time", "price of passage">,

Φ_{11} - <"emergency situation">,

Φ_{12} - <"direction of transportation">,

Φ_{13} - <"direction of transportation", "time", "price of the passage">.

As with E_c, components which do not participate directly in the model will be omitted. This GN is a reduced one from the same class of reduced GNs as the first one.

The transition

$Z_1 = \langle \{l_6, l_8, l_{14}\}, \{l_1\}, r_1, v(l_6, l_8, l_{14})\rangle$,

where

$$r_1 = \begin{array}{c|c} & l_1 \\ \hline l_6 & true \\ l_8 & true \\ l_{14} & true \end{array}$$

denotes "the transportation means is ready for taking up goods/out of

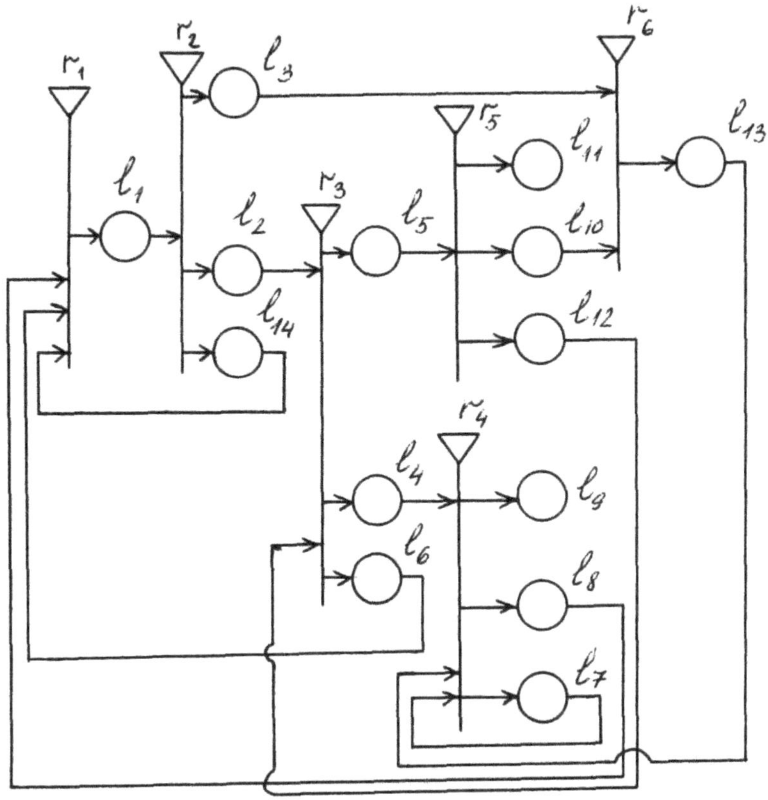

Fig. 3.2

usage".

$Z_2 = \langle \{l_1\}, \{l_2, l_3, l_{14}\}, r_2, \vee(l_1)\rangle,$

where

$$r_2 = \begin{array}{c|ccc} & l_2 & l_3 & l_{14} \\ \hline l_1 & W_1 & W_2 & W_3 \end{array}$$

where

W_1 = "the transportation means is working",

W_2 = "the transportation means is sent to meet the goods",

W_3 = "the transportation means is sent to be repaired".

Z_2 denotes "the transportation means is sent to meet the goods".

$Z_3 = \langle\{l_2, l_{12}\}, \{l_4, l_5, l_6\}, r_3, \vee(l_2, l_{12})\rangle$,

where

$$r_3 = \begin{array}{c|ccc} & l_4 & l_5 & l_6 \\ \hline l_2 & W_4 & W_5 & W_6 \\ l_{12} & false & true & false \end{array},$$

where

W_4 = "the transportation means is moving down a road",

W_5 = "the transportation means is on water/railway",

W_6 = "leaving of goods and moving back for taking the new ones".

$Z_4 = \langle\{l_4, l_7, l_{13}\}, \{l_7, l_8, l_9\}, r_4, \vee(l_4, l_7, l_{13})\rangle$,

where

$$r_4 = \begin{array}{c|ccc} & l_7 & l_8 & l_9 \\ \hline l_4 & W_6 & W_7 & W_8 \\ l_7 & W_6 & W_7 & W_8 \\ l_{13} & W_6 & W_7 & W_8 \end{array},$$

where

W_6 = "the transportation means is moving on",

W_7 = "the transportation means is at a new location",

W_8 = "the transportation means is having an emergency".

Z_4 denotes "the goods are transported by land to the consumer". That transition aims to model the movement of the transportation means by land in different countries.

$Z_5 = \langle\{l_5\}, \{l_{10}, l_{11}, l_{12}\}, r_5, \vee(l_5)\rangle,$

where

$$r_5 = \begin{array}{c|ccc} & l_{10} & l_{11} & l_{12} \\ \hline l_5 & W_9 & W_{10} & W_{11} \end{array},$$

where

W_9 = "trailer/road railer at the end of the second stage of transportation";

W_{10} = "trailer/road railer having emergency";

W_{11} = "trailer/road railer moving on".

Z_5 denotes "movement of the transportation means in the second stage of multimodal transportation".

$Z_6 = \langle\{l_9, l_{10}\}, \{l_{13}\}, r_5, \wedge(l_9, l_{10})\rangle,$

where

$$r_6 = \begin{array}{c|c} & l_{13} \\ \hline l_9 & true \\ & \\ l_{10} & true \end{array}.$$

The transition Z_6 is of the $\wedge$-type because it shows the existence of at least one token at both its input places. It describes the "end of the second stage of multimodal transportation and begining of a third one".

Unlike GN E_c, there is a cycle between the transitions Z_5 and

Z_3 here, which corresponds to going in a certain direction. The notion of the transportation means with a change of their locations in universe causes increasing charges for passing to a definite stage and leads to changes in the characteristics of the tokens. The GN E_n can be used for

(a) prognosing the expenses of the transportation firm for giving service when knowing the direction and the way of changing the characteristics;

(b) if the time-parameters related to the tokens are known, the transportation firm can follow their movement and send them for disposal of the forwarding agent for a further use and new contracts.

The above models show only a part of the possibilities of GNs. These models could be used not only for simulation of concrete processes, but also for management (in combination with expert systems) and control in real time.

This paper is pased on [1].

REFERENCE:

[1] A. Kirova, K. Kirov, K. Atanassov, Modelling the international transporting market activities with generalized nets "Avtomatiua, Iztchislitelna Tehnika i Avtomatizirani sistemi", 1985, No. 6, 61-66. (in Bulgarian).

§ 3.2: GENERALIZED NETS FOR SIMULATION OF BOOKING SYSTEMS FOR PASSENGER TRANSPORT

Sergej Nedev and Lilija Atanassova

The automated systems for booking and sale of tickets are territorially distributed complexes having a set of parallel processes. The building of such systems requires great investment. Mathematical models are necessary for a detailed analysis of the parameters and function of the complexes, and for correcting errors in the project itself. The result of this analysis can be used for choosing an optimal hardware and software for building of the system.

A mathematical instrument for analysing such systems is the simulation based on GNs. A GN represents not only the structure of the process but also the dynamics of the process. In comparison with other net models, GNs are very compact and simple in structure. They provide very detailed information about the simulated processes.

In [1] a model of a booking system is described. Here the topic of the simulated processes will be extended. Let us examine the GN model of a fragment of a booking system. The fragment consists of the following components:

1. Points for booking - 30;
2. Box-offices at every point, numbered from 1 to 8;
3. The main computer centre of the system (MCC);
4. Telephone channels for communication between the MCC and the points (three logic channels, each containing 5 physical channels).

The model gives the possibility of the information about a part of the vehicles to be stored at the points from which they start. The channel access is of the type CSMA/CD. The maximum length of a message is 20 bytes. The transmission speeds in the channels are:

MCC-point - 2400 bit/s;

point-box-office - 500 000 bit/s.

MCC response to a request is 0.5s.

Our GN model is based on the following reduced GN (see § 1.1)

$$\langle A, \langle K, \theta\rangle, \langle X, \Phi\rangle\rangle,$$

where

$A = \{Z_1, \ldots, Z_9\}$ is a set of transitions of GN, K is a set of tokens. Every system's request is represented by exactly one token in the GN model. Every token has the initial characteristic of a couple of position integers (m, n), where m is the number of the booking point and n the number of the box-office at the booking point. These token characteristics are exactly the elements of the set X. The function θ specifies for every token $\alpha\in K$ the moment of time it will enter the GN through the initial place l_1. It is convenient to considered the characteristic function as

$$\Phi = \bigcup_{i=2}^{16} \Phi_i,$$

where Φ_i is a characteristic function related to the place l_i (the tokens enter l_1 with their previous characteristics without receiving such in l_1). A variable can be specified in the model. Its values correspond to the increment of time for the simulated process. This component is not related to the global time components of the GN. The global time components are not necessary for this model and that is why they are omitted. Let T denote the current value of this variable, T_0 is its initial value, t_α is the time for the stay of the token α in the GN and t_i the time for the stay of a token in the place l_i ($1 \le i \le 16$).

The places and the characteristic functions related to them have the following meanings:

l_1 - stands for box-offices. In the box-offices the users requests are are created, shown as tokens of the GN.

l_2 - stands for points of booking. All requests are collected at the point.

Φ_2 - specifies the value $X_\alpha + t_1$ as a characteristic of the tokens, where X_α is the final characteristic of the token α and t_1 = 0.002 s . K, where K is a random number for which $1 \le K \le 9$ (the number 0.002s is the duration of a single transmission between a box-office and a booking point).

l_3 - stands for the queue of requests, waiting for sending through the telephone channel between the booking point and the MCC.

Φ_3 - specifies the duration of the tokens' stay in place l_3.

For $4 \le i \le 8$:

l_i - stands for sending of the request through one of the telephone channels

Φ_i - specifies the value $X_\alpha + t_2 + t_3$ as a characteristic of the tokens, where $t_2 = 0.1 \cdot K$, where K is a random number, $0 \le K \le 4$. It is assumed that in 70% of the cases are in $0 \le K \le 1$. The number 0.1 is the duration in seconds of the sending of a request through the telephone channel, $t_3 = 0.1 \cdot (Q + 1)$, where Q is the number of the simultaneously transmitted requests $(4 \le i \le 8)$.

l_9 - stands for request reception in the MCC of the system

Φ_9 - specifies the duration of the tokens' stay in the place l_9.

l_{10} - stands for request processing in the MCC.

Φ_{10} - specifies the value $X_\alpha + t_4$ as a characteristic of the tokens, where $t_4 = 0.5$ s. (It is assumed that one request is processed in the average time of 0.5 seconds).

l_{11} - stands for the queue of requests, processed in the booking point itself.

Φ_{11} - specifies the duration of the token's stay in the place l_{11}.

l_{12} - stands for request processing in the booking point.

$\Phi_{12} = \Phi_{10}$.

l_{13} - stands for cashier reception of the system response.

Φ_{13} - specifies the value $X_{\alpha}+t_1+t_5$ as a characteristic of the tokens, where $t_5 = L$, L being a random number equally distributed in the the interval 60 - 120s.

l_{14} - stands for the end of request processing.

l_{15} - is analogous to l_{13}.

$\Phi_{15} = \Phi_{13}$.

l_{16} - is analogous to l_{14}.

The capacities of the different places are: for l_1 - infinite, for l_2, l_3, l_9, l_{11}, l_{13}, l_{14}, l_{15}, l_{16} - 240, for l_{12} - 30, for l_{10} - 1, for l_4, l_5, l_6, l_7, l_8 - 24. The capacity of each arc of the transitions is 1, e.g. every element of the capacity matrix is equal to 1 and that is why this component will be omitted in the transition definition below. Besides all the transitions are disjunctive and that is why this component will be omitted too.

The formal description of the GN transitions of Fig. 3.3 (bearing in mind that this GN is reduced) is

$$Z_1 = \langle \{l_1\}, \{l_2\}, r_1 \rangle,$$

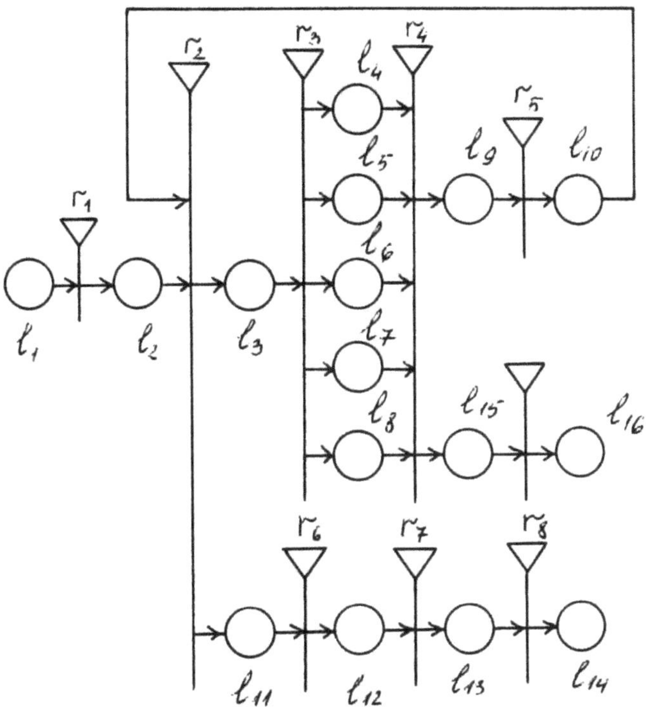

Fig. 3.3

where l_1 is the input place for Z_1 and l_2 is the output place for Z_1, the transition condition is

$$r_1 = \begin{array}{c|c} & l_2 \\ \hline l_1 & W_1 \end{array},$$

where

W_1 = "in the GN does not exist any other token with the same initial characteristic".

$$Z_2 = \langle \{l_2, l_{10}\}, \{l_3, l_{11}\}, r_2 \rangle,$$

where

$$r_2 = \begin{array}{c|cc} & l_3 & l_{11} \\ \hline l_2 & W_2 & W_3 \\ l_{10} & W_4 & false \end{array}$$

where

W_2 = "$\partial = 0$" & "$T \geq X_\alpha + t_1$",

W_3 = "$\partial = 1$" & "$T \geq X_\alpha + t_1$",

where ∂ is an indication for branching at the transition Z_2.

W_3 = "$T \geq X_\alpha + t_4$".

$$Z_3 = \langle \{l_3\}, \{l_4, l_5, l_6, l_7, l_8\}, r_3 \rangle,$$

where the r_3's predicates contain the essential of the channel access type CSMA/ CD.

$$Z_4 = \langle \{l_4, l_5, l_6, l_7, l_8\}, \{l_9, l_{15}\}, r_4 \rangle,$$

where

$$r_4 = \begin{array}{c|cc} & l_9 & l_{15} \\ \hline l_4 & W_4 & W_5 \\ l_5 & W_4 & W_5 \\ l_6 & W_4 & W_5 \\ l_7 & W_4 & W_5 \\ l_8 & W_4 & W_5 \end{array}$$

where

W_4 = "$\sigma = 0$" & "$T \geq X_\alpha + t_2 + t_3$",

W_5 = "$\sigma = 1$" & "$T \geq X_\alpha + t_2 + t_3$",

where σ is an indication for branching at the transition Z_4.

$$Z_5 = \langle \{l_9\}, \{l_{10}\}, r_5 \rangle,$$

where

$$r_5 = \begin{array}{c|c} & l_{10} \\ \hline l_9 & true \end{array};$$

$$Z_6 = \langle \{l_{11}\}, \{l_{12}\}, r_6 \rangle$$

where

$$r_6 = \begin{array}{c|c} & l_{12} \\ \hline l_{11} & W_6 \end{array}$$

where

W_6 = "in l_{12} there are no tokens with the same characteristic m";

$$Z_7 = \langle \{l_{12}\}, \{l_{13}\}, r_7 \rangle,$$

where

$$r_7 = \begin{array}{c|c} & l_{13} \\ \hline l_{12} & W_7 \end{array}$$

where

W_7 = " $T \geq X_\alpha + t_4$ ";

$$Z_8 = \langle \{l_{13}\}, \{l_{14}\}, r_8 \rangle,$$

where

$$r_8 = \begin{array}{c|c} & l_{14} \\ \hline l_{13} & W_8 \end{array}$$

where

W_8 = "$T \geq t_1 + t_5 + X_\alpha$".

$$Z_9 = \langle \{l_{15}\}, \{l_{16}\}, r_9 \rangle$$

where

$$r_8 = \begin{array}{c|c} & l_{16} \\ \hline l_{15} & W_8 \end{array} .$$

The described model shows the basic points of the structure and the function of an automated system for booking and sale of tickets in the passenger transport. This model can be used for two principal aims:

1. When the main parameters of the system are specified such as productivity of the MCC, loading of the system, transmission capacity of the channels, by this model the following characteristics of the system can be defined:

a) average time for processing of a request;

b) maximum time for processing of a request;

c) the part of processed requests for a fixed period of time;

d) the time for processing the requests in every element of the system;

2. When the loading of the system is specified, the productivity of the elements can be defined so that the average time for processing of a request does not overrun a certain value.

3. The number of the processed requests for a defined period of time and others.

The suggested model can be developed, if necessary, for getting extra information about the simulated processes.

This paper is based on [2].

REFERENCES:

[1] S. Nedev, K. Atanassov, An automated booking system modellng by generalized networks. Automation, Computer Engineering and Automa-

tion Systems, 1989, No. 6, 17-20 (in Bulgarian).

[2] S. Nedev, L. Atanassova, Generalized nets for simulation of rezervation systems for passenger transport, First Sci. Session of the Math. Found. AI Seminar, Sofia, Oct. 10, 1989, Preprint IM-MFAIS-7-89, Sofia, 1989, 45-49.

§ 3.3: GENERALIZED NETS IN MODELLING OF PUBLIC TRANSPORT

Katja Stefanova

In this section, public transport is modelled by means of the Generalized Nets (GNs).

"A transport system" means the set consisting of a transport net in a given urbanized region, all vehicles and passengers, and other active or passive participants in the public traffic of the same region.

A transport net is characterized by its geometry, topology and organization. Its aim is to realise vehicle or pedestrian connection between all points which are to be connected in a given region.

We shall describe transport nets involving public transport (buses, trams, trolley-buses etc.) only. Such a transport net can be characterized by a directed graph G whose vertices represent the stations. Each pair of vertices representing neighbouring stations of a given route is connected with arcs directed according to the traffic.

In the sequel we shall use the GN for transport systems modelling.

Let

$$V = \{v_1, v_2, \ldots, v_p\}$$

be the set of vertices, and

$$A = \{a_1, a_2, \ldots, a_q\}$$

be the set of arcs of the graph G. To each vertex $v \in V$ and its corresponding input and output arcs $a'_1, a'_2, \ldots, a'_i; a''_1, a''_2, \ldots, a''_j$, where $1 \le i \le p$, $1 \le j \le q$ (see Fig. 3.4), we shall assign the transitions Z' and Z'' from the GN (see Fig. 3.5).

In view of the results of § 5.1 in App. 1, the union of all pairs of transitions corresponding to the vertices of the transport

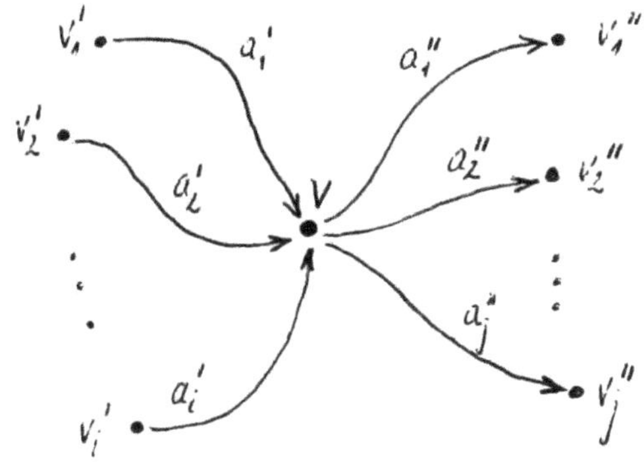

Fig. 3.4

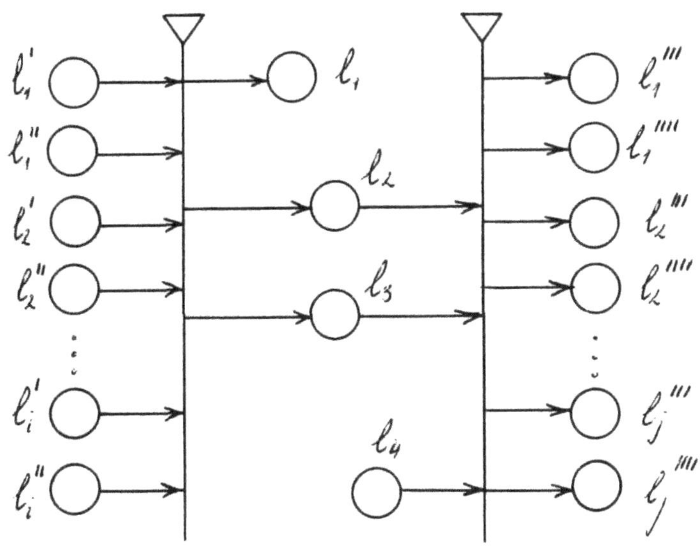

Fig. 3.5

net represents a GN. This GN contains transitions of the two types given above and two kinds of tokens corresponding to the vehicles and the passenger stream. The tokens of the first kind will run over places of the type l', l'' and l_2 and the second kind will run over places of types l''', l'''', l_1, l_2, l_3.

The GN will also contain global time parameters such as an initial moment T, an elementary time-step t^o (e.g. a minute, for "rough" models the time step may be larger), duration of the model functioning t^* (hours up to twenty-four hours). The first kind of tokens of the GN has the following initial characteristics: type of the vehicles, registration numbers, numbers of the line. As for the second kind of token, the number of the passengers boarding a vehicle is its initial characteristic. The tokens of the first kind are in the GN at the moment T, whereas the tokens of the second kind are in the GN during its functioning through the places of type l_4. In the traffic, the tokens of the first kind can be characterized also by "the time necessary for a transition from an ex-station to the current station" (in places of type l') as well as by "the sojourn time in the last station" (in a place of type l''). The tokens of the second kind can be characterized by "the number of passengers who are currently in the vehicle" (in a place of type l'''') and "the number of passengers getting off the vehicle" (in a place of type l_1). The tokens of the second kind do not possess any characteristics in a place of type l_3.

A transition of type Z' has the following form:

$$Z' = \langle\{l'_1, l'_2, \ldots, l'_i, l''_1, l''_2, \ldots, l''_i\}, \{l_1, l_2, l_3\}, *, *, r', *,$$
$$\vee(\wedge(l'_1, l''_1), \wedge(l'_2, l''_2), \ldots, \wedge(l'_i, l''_i))\rangle,$$

where

a) $l'_1, l'_2, \ldots, l'_i$ and $l''_1, l''_2, \ldots, l''_i$ correspond to the arcs between the vertices of the graph which are connected with the vertex v;

b) l_1 is related to passengers getting off the vehicle (the tokens receive in it the number of these passengers);

c) l_2 is related to the presence of the vehicle;

d) l_3 is related to the presence of passengers in the vehicle (the tokens receive in it the number of these passengers);

$r' =$

	l_1	l_2	l_3
l'_1	false	true	false
$\vdots$	$\vdots$	$\vdots$	$\vdots$
l'_i	false	true	false
l''_1	W_1	false	W_2
$\vdots$	$\vdots$	$\vdots$	$\vdots$
l''_j	W_1	false	W_2

where

W_1 = "there is at least one passenger getting off the vehicle";

W_2 = "there is at least one passenger continuing the trip".

A transitions of type Z" has the following form:

$$Z'' = \langle \{l_2, l_3, l_4\}, \{l'''_1, l'''_2, \ldots, l'''_i, l''''_1, l''''_2, \ldots, l''''_i\}, *, *, r'', *, \wedge(l_2, \vee(l_3, l_4))\rangle,$$

where

a) place l_4 is related to passengers boarding the vehicle (the tokens receive in it the number of these passengers);

e) the transition condition is

$$r'' = \begin{array}{c|cccccc} & l'''_1 & \ldots & l'''_j & l''''_1 & \ldots & l''''_j \\ \hline l_2 & W_3 & \ldots & W_3 & false & \ldots & false \\ l_3 & false & \ldots & false & W_3 & \ldots & W_3 \\ l_4 & false & \ldots & false & W_3 & \ldots & W_3 \end{array}$$

where

W_3 = "the next route station of the vehicle".

All transitions have equal priorities.

The GN described above is a reduced GN of the type

$$\Sigma^{A_3, A_4, A_6, \pi_A, \theta_1, \theta_2, b}.$$

We conclude that within the framework of the model we can specify the following main characteristics: priority of the stations (for the places of type l_2), capacity of the places of type l_2 (number of vehicles stopping at a given station) and the priority of the vehicles (tokens of the first kind).

The formullation of an analytical model of the considered transport system, particularly the analytical part of the given GN model, is related to the influence of factors of various nature. Some of them are: settings and movements of the population, presence of different social and age groups with the appropriate requirements for security, comfort, costs, and travel time. These factors affect the determination of the initial parameters of the second kind of tokens (number of passengers waiting for a vehicle at a given station, with preferences regarding the type of vehicles, current number of passengers, etc.).

Different characteristics of the transport system can be obtained by simulating the model built by means of a GN. It makes it possible to use the GN model as a tool for transport process management. Furthermore, it enables optimization of the vehicle routes simulating the traffic process and passenger streams. It is easy to obtain the average and maximum time duration of a given route. In the case of any changes, the time schedule and the correspondingly adjusted passenger streams can be fixed. On the basis of the obtained results, stratification of the transport net and appropriate estimation of the travel costs can be determined.

§ 3.4: GENERALIZED NET MODEL OF ACTIVITIES OF A BUS-DEPOT

Pavlin Gyurov

An example for using the Generalized Nets (GNs) for a description of some transport situations is introduced. It illustrates the solving of the following problem.

Suppose a public coach agency uses n coaches whose numbers of seats are known. When the coaches are not in use, they are in a garage. There is a service for maintenance work. The coaches leave for the routes from a bus station. Every morning, before leaving the garage, they take their own daily time table. At the end of their workday, they go back to the garage.

This process will be described below by means of a GN and the loading of the lines and coaches will be reported.

The GN has a graphical structure as shown in Fig. 3.6.

Every coach is represented by a token in the GN. When a coach is on a route, the token corresponding to it goes into the GN. At the time-moment T, all coaches are in the garage, so at that moment all bus-tokens BT_i $(1 \le i \le n)$ are in place l_3. If any coach is out of order, it has to be repaired so the corresponding token goes to place l_{13} through place l_{15} and the transitions Z_2 and Z_3. When the coach is in order again the corresponding token will come back to place l_3 through place l_1 and the transitions Z_4 and Z_1. If any coach located in the garage is in order and it is time to go on a route, the corresponding token leaves place l_3 and goes to place l_7 through place l_5 and the transitions Z_7 and Z_5. If there is much time before the next travel of the coach, it goes to the garage and the corresponding token

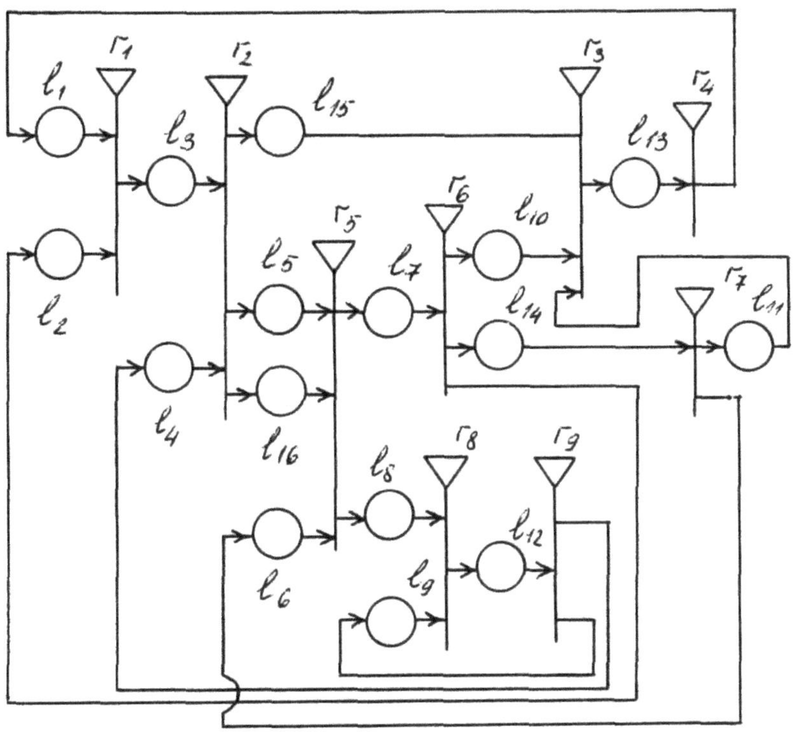

Fig. 3.6

goes to place l_3 through place l_2 and the transitions Z_6 and Z_1. The situation is the same when its travel is completed.

However, if a coach is damaged and has to go on the route, or it gets damaged while travelling, it cannot do the next planned travel. So it has to go for servicing. Then the corresponding token goes to place l_{13}. But the travelling will not be done. So it is necessary that another coach takes up the next travel. This is why there is a private token in the GN. At the time-moment T it is in place l_{12}. This token (which we will mark below by OT) gets as characteristic the next travel which the damaged coach would to do. Then it goes to place l_4. In this case, there is a choice for an optimal coach from those in the garage. The corresponding token goes to place l_5 and the goes to place l_{16}. The bus-token goes to place l_7 and OT comes back to place l_{12} through place l_8 and the transitions Z_5 and Z_8. If there is no coach that can accept the order, OT waits in place l_4. It is possible that at that time-moment another coach come out of order. So OT splits in the transition Z_9 and its copy comes back in place l_{12} through place l_9 and the transition Z_8. It is useful if the transitions Z_8 and Z_9 work more often than the others.

Bus-tokens have only one (current) characteristic with the form: "<seats, maintenance, travels>", where

seats is a number of sites,

maintenance = <duration of maintenance, repairs>,

where repairs is the set of repairs of the coach up to the moment,

travels = $\{<direction(t), passengers(t), duration(t)> / t \in [T, T+t^*]\}$

OT can have a characteristic of the form (see above)

"<seats, waiting-time, travels">.

Initially, it is in place l_{12}. When it comes into place l_4, it gets a new characteristic whose seats component is the number of sites of the damaged coach, its waiting-time component is the previous one and the travels component is the set of travels which this coach would do. When OT comes into place l_9, it doesn't change its characteristic. In place l_{16} it gets new characteristics - the previous seats and travel components and the waiting-time characteristic is the duration of is stay in place l_4.

Every bus-token gets as a first characteristic its number of seats and its daily time table. The maintenance component of its first characteristic is $\langle 0, \emptyset \rangle$. When it goes to place l_{13} it changes its characteristic. Travel component reduces to the set of travel fulfilled. The current characteristic of the token includes information about it. When a token comes in place l_1 it gets a characteristic - the previous travel component. The number of seats can be changed. The last of the repairs join the repair component, and the duration of these repairs adds to the current duration of maintenance.

When a token comes into place l_{14} it changes the passengers (TIME) characteristic. A bus-token that comes into place l_3 can append new travels to the travel component of its characteristic.

$$Z_1 = \langle \{l_1, l_2\}, \{l_3\}, t_1^1, t_2^1, r_1, M_1, \vee(l_1, l_2) \rangle,$$

where

$$r_1 = \begin{array}{c|c} & l_3 \\ \hline l_1 & true \\ l_2 & true \end{array}$$

$$M_1 = \begin{array}{c|c} & l_3 \\ \hline l_1 & n \\ l_2 & n \end{array},$$

$Z_2 = \langle\{l_3, l_4\}, \{l_{15}, l_5, l_{16}\}, t_1^2, t_2^2, r_2, M_2, \vee(l_3, \wedge(l_3, l_4))\rangle,$

where

$$r_2 = \begin{array}{c|ccc} & l_{15} & l_5 & l_{16} \\ \hline l_3 & W_1 & W_2 & false \\ l_4 & false & false & W_3 \end{array}$$

where

W_1 = "the coach is out of order";

W_2 = $\neg W_1$ & ("the moment for going to the station has come" or ("there is an OT in l_4" and ("the coach has no other travel" or "the travel of OT can be done before the travel of the bus-token") and "this coach is optimal about the number of seats");

W_3 = "the order is taken up by the bus-token";

$$M_2 = \begin{array}{c|cc} & l_{15} & l_5 \\ \hline l_3 & n & n \\ l_4 & 0 & 1 \end{array}$$

$Z_3 = \langle\{l_{15}, l_{10}, l_{11}\}, \{l_{13}\}, t_1^3, t_2^3, r_3, M_3, \vee(l_{15}, l_{10}, l_{11})\rangle,$

where

$$r_3 = \begin{array}{c|c} & l_{13} \\ \hline l_{15} & true \\ l_{10} & true \\ l_{11} & true \end{array}$$

$M_3 =$

	l_{13}
l_{15}	n
l_{10}	n
l_{11}	n

$$Z_4 = \langle \{l_{13}\}, \{l_1\}, t_1^4, t_2^4, r_4, M_4, v(l_{13})\rangle,$$

where

$r_4 =$

	l_1
l_{13}	true

$M_4 =$

	l_{13}
l_{13}	n

$$Z_5 = \langle \{l_5, l_6, l_{16}\}, \{l_7, l_8\}, t_1^5, t_2^5, r_5, M_5, v(l_5, l_6, l_{16})\rangle,$$

where

$r_5 =$

	l_7	l_8
l_5	true	false
l_6	true	false
l_{16}	false	true

$M_5 =$

	l_7	l_8
l_5	n	0
l_6	n	0
l_{16}	0	1

$Z_6 = \langle\{l_7\}, \{l_{10}, l_2, l_{11}\}, t_1^6, t_2^6, r_6, M_6, v(l_7)\rangle,$

where

$$r_6 = \begin{array}{c|ccc} & l_{10} & l_{14} & l_2 \\ \hline l_7 & W_1 & W_4 & W_5 \end{array}$$

where

$W_4 = \neg W_1$ & "it is time corresponding to the token coach to travel",

$W_5 = \neg W_1$ & ("there is much time before the next travel" or "the coach's work day is finished"),

$$M_6 = \begin{array}{c|ccc} & l_{10} & l_{14} & l_2 \\ \hline l_7 & n & n & n \end{array};$$

$Z_7 = \langle\{l_{14}\}, \{l_{11}, l_6\}, t_1^7, t_2^7, r_7, M_7, v(l_{14})\rangle,$

where

$$r_7 = \begin{array}{c|cc} & l_{11} & l_6 \\ \hline l_{14} & W_1 & W_6 \end{array}$$

where

$W_6 = \neg W_1$ & "it is the end of travel",

$$M_7 = \begin{array}{c|cc} & l_{11} & l_6 \\ \hline l_{14} & n & n \end{array};$$

$Z_8 = \langle\{l_8, l_9\}, \{l_{12}\}, t_1^8, t_2^8, r_8, M_8, v(l_8, l_9)\rangle,$

where

$$r_8 = \begin{array}{c|c} & l_{12} \\ \hline l_8 & true \\ l_9 & true \end{array}$$

$$M_8 = \begin{array}{c|c} & l_{12} \\ \hline l_8 & 1 \\ l_9 & 1 \end{array},$$

$$Z_9 = \langle \{l_{12}\}, \{l_2, l_9\}, t_1^9, t_2^9, r_9, M_9, \vee(l_{12}) \rangle,$$

where

$$r_9 = \begin{array}{c|cc} & l_2 & l_9 \\ \hline l_{12} & W_7 & true \end{array}$$

where

W_7 = "there is a coach which is damaged" & "this coach has to travels",

$$M_9 = \begin{array}{c|cc} & l_2 & l_9 \\ \hline l_{12} & 1 & 1 \end{array}.$$

For the GN-components to be valid,

$\pi_A(Z_i) = 1$ for $1 \le i \le 9$,

$\pi_L(l_4) = \pi(l_9) = 1$; for all other places: $\pi_L(l) = 2$,

$c(l_{12}) = 2$; for all other places: $c(l) = n$,

$$\theta_1^i(t_1^i) = \begin{cases} t_1^i + 2.t^o, & \text{if } i \in \{1, 2, 3, 5, 6, 7\} \\ t_1^i + C.t^o, & \text{if } i = 4 \ (C = \text{const} \geq 2), \\ t_1^i + t^o, & \text{if } i \in \{8, 9\} \end{cases}$$

It means that the private tokens go faster than others. In order to maintain the works character it is possible that the transition Z_4 works rarely.

$$\theta_2^i(t_1^i) = t_2^i = t^o \quad (1 \leq i \leq 9).$$

The numbers t_2^i $(1 \leq i \leq 9)$ must be large enough so that all tokens can go through transitions

$K = \{BT_i \ / \ 1 \leq i \leq n\} \cup \{OT\}$,

$\pi_K(\alpha) = 1$, for every $\alpha \in K$,

$\theta_K(\alpha) = T$, for every $\alpha \in K$.

All tokens enter the GN at the starting time-moment of its work. T is the time-moment of the start of workday, $t^o = 0.5$ minutes (for example); $T + t^*$ is the time-moment of end of a workday.

§ 3.5: GENERALIZED NET MODEL OF THE ACTIVITY OF AN AIR-TRAFFIC CONTROL CENTER

Stanislav Pishmanov

As an example of the description of real processes with Generalized Nets (GNs) we shall examine a GN-model of the activity of an air-traffic control center at an airport. In general, this activity consists of the following: when an airplane is coming to the airport area for landing, it comes under the control of a dispatcher, who fixes an air-corridor where the airplane can fly a round until a runway is free. At the same time, there must also be air-corridors for airplanes which have just flown off until they leave the airport area. We shall examine an airport with n tracks (each one of which can be used for both landing and for flying off), where the air-traffic is being controlled by 4 dispatchers - one for each direction: North (1), West (2), South (3) and East (4). All tracks are used by all dispatchers at the same time. Every dispatcher can control up to m airplanes at any one moment, while in special situations for short intervals of time this number can be increased to $m + k$, where k is a fixed natural number, i.e. the normal capacity of the airport is $4.m$ airplanes waiting for landing at any one moment. When the number of waiting airplanes is less than $4.m$, this means that there are free air-corridors. We shall examine the airplanes which have just flown off from the same way as those waiting for landing - each takes one air-corridor from the moment of flying off till the moment of leaving the airport area.

Let us examine a GN with 7 transitions and 23 places for movement of airplanes, and a separate transition Z_0 with 2 places, whit tokens moving, carrying managing information (see Fig. 3.7).

The separate transition Z_0 has two input places l_0 and l_1 which are also the output places for it. In each of them, there is a token which passes the transition; when it is activated, it renews its transition and goes back to its starting place, i.e. the index matrix of the transition Z_0 is

	l_0	l_1
l_0	true	false
l_1	false	true

.

There is only one token α with the current characteristic (and without the other ones) x^{α} = "$\langle a, a_1, a_2, \ldots, a_n \rangle$", where a is the number of current free air-corridors of the whole airport at the moment, $a_1, a_2, \ldots, a_n \in \{0, 1\}$, $a_i = 0$ means that the i-th track is occupied at the moment, and $a_i = 1$ means the opposite situation ($1 \le i \le n$). In the place l_1 there is only one token - β, which has the characteristic x^{β} = "$\langle b_1, b_2, b_3, b_4 \rangle$", where b_i is a natural number for which $-k \le b_i \le m$, and b_i is the number of free air-corridors under the control of the i-th dispatcher ($1 \le i \le 4$).

The set S with elements s_j is the set of tokens which represent the airplanes in the airport area. Their number is an element of the set $\{1, 2, \ldots, 4.m\}$ and they are transferred to the other section of the GN. Each of these tokens has as an initial characteristic its identifying data (type, nationality, number of course, etc.). In the place l_4 receives as new characteristic "$\langle c^j_1, c^j_2, c^j_3 \rangle$" ($1 \le i \le 4.m$), where:

- $c_1^j \in \{1, 2, 3, 4\}$ determines the direction from which the airplane comes; for airplanes which have just flown off, this characteristic determines the direction to which the airplane will be travelling; inside the GN, for airplanes waiting to land, this characteristic carries the appropriate information;
- c_2^j is a parameter corresponding to the time during which the airplane can fly with its fuel; for the airplanes just taking off it always has the maximum value;
- $c_3^j \in \{0, 1\}$ and it is equal to 0 for an airplane waiting to land and to 1 for an airplane just taking off.

Token s_j $(1 \le j \le 4.m)$ can enter the GN through three input places l_1, l_2 and l_{16}. Every token entering through l_2 corresponds to the airplane, coming to the airport area for landing. If the airport capacity is not exceeding (i.e. $pr_1 x^\alpha \neq 0$), then the token moves to place l_4 (where it receives the above characteristic) and (simultaneously the token α receives as a new characteristic such one with $pr_1 x^\alpha$ decreased by 1 (the airplane is accepted in the airport). Otherwise, the token moves to place l_3, which is an output place for tokens corresponding to airplanes (the airplane is directed to another airport). The transition condition of Z_1 is

$$r_1 = \begin{array}{c|cc} & l_3 & l_4 \\ \hline l_2 & W_1 & W_2 \end{array}$$

where

$W_2 = $ "$pr_1 x^{\alpha} \neq 0$",

$W_1 = \neg W_2$.

A token s_j which has entered the GN through the second input place (l_{15}), corresponds to an airplane just taking off and has $c_3^j = 1$. To move through the transition Z_3, it is necessary to satisfy the condition "$pr_{2,3,\ldots,n+1} x^{\alpha} \neq \langle 0, 0, \ldots, 0 \rangle$" & "$pr_1 x^{\alpha} \neq 0$" (i.e. at least one track must be free and also at least one air-corridor for leaving the airport must be free). Besides a token leaving place l_{15} can move only to place l_{16} for assignment to a dispatcher, i.e. the row of the index matrix of the transition Z_3 corresponding to place l_{15} contains the value "false" in every position, so does the column corresponding to place l_{16}, with only one exception: the predicate related to the places l_{15} and l_{16} is "$pr_{2,3,\ldots,n+1} x^{\alpha} \neq \langle 0, 0, \ldots, 0 \rangle$" & "$pr_1 x^{\alpha} \neq 0$". When a token moves from place l_{15} to place l_{16}, the token α receives a new characteristic: $pr_1 x^{\alpha}$ decreases by 1.

The transition Z_2 stands for the distribution of airplanes to the dispatcher. A token s_j moving through this transition, regardless of the place where it is coming from, goes to place l_{4+i} (the corresponding airplane is distributed to the i-th dispatcher), where $i = c_1^j$. The transition condition of Z_2 is

$$r_2 = \begin{array}{c|cccc} & l_5 & l_6 & l_7 & l_8 \\ \hline l_3 & W_3 & W_4 & W_5 & W_6 \\ l_{17} & W_3 & W_4 & W_5 & W_6 \\ l_{18} & W_3 & W_4 & W_5 & W_6 \\ l_{19} & W_3 & W_4 & W_5 & W_6 \\ l_{20} & W_3 & W_4 & W_5 & W_6 \\ l_{16} & W_3 & W_4 & W_5 & W_6 \end{array}$$

where

W_{2+i} = "$c_1^j = i$" for $1 \le i \le 4$.

Transition Z_3 contains two output (of the GN) places - l_{13} and l_{14}. Tokens, moving to place l_{14} correspond to airplanes just flown off, which are leaving the airport area under the dispatcher's guidance. A token s_j can move to place l_{14}, if it comes from place l_{4+i} for $1 \le i \le 4$, and if $c_2^j = 1$. Then $pr_1 x^\alpha$ and $pr_i x^\beta$ each increases by 1. Tokens moving to the other output place l_{13} correspond to airplanes, which are landing at the moment. A token s_j can move to place l_{13} if if it has $c_3^j = 0$ and "$pr_{2, 3, \ldots, n+1} x^\alpha \ne \langle 0, 0, \ldots, 0\rangle$" (at least one track is free). After this, $pr_1 x^\alpha$ and $pr_i x^\beta$ each increase by 1, where i depends on the GN-place where the token comes from

$$i = \begin{cases} j - 4, & \text{for } 5 \le j \le 8 \\ j - 20, & \text{for } 21 \le j \le 24 \end{cases}$$

l_7 and l_8 respectively, with the exception that in places l_{21}, l_{22}, l_{23} and l_{24} tokens corresponding to airplanes just taking off cannot appear. If the equation $c_3^j = 0$ (airplane waiting for landing) is valid for a token s_j, it comes from place l_{4+i} $(1 \le i \le 4)$ or place l_{20+i} $(1 \le i \le 4)$, and $pr_{2,3,\ldots,n+1} x^{\alpha} = \langle 0, 0, \ldots, 0\rangle$ at this moment, it moves to place l_{8+i} (the corresponding airplane receives a free air-corridor for waiting for a track to be free, the airplane staying under the control of the same dispatcher). From the above, it follows that the transition condition r_3 has the form

r_3 =

	l_9	l_{10}	l_{11}	l_{12}	l_{13}	l_{14}	l_{16}
l_5	W_7	false	false	false	W_8	W_9	false
l_6	false	W_7	false	false	W_8	W_9	false
l_7	false	false	W_7	false	W_8	W_9	false
l_8	false	false	false	W_7	W_8	W_9	false
l_{21}	W_7	false	false	false	W_8	W_9	false
l_{22}	false	W_7	false	false	W_8	W_9	false
l_{23}	false	false	W_7	false	W_8	W_9	false
l_{24}	false	false	false	W_7	W_8	W_9	false
l_{15}	false	false	false	false	false	false	W_{10}

where

W_7 = "$pr_{2,3,\ldots,n+1} x^{\alpha} = \langle 0, 0, \ldots, 0\rangle$" & "$c_2^j = 0$",

W_8 = "$pr_{2,3,\ldots,n+1} x^{\alpha} \neq \langle 0, 0, \ldots, 0\rangle$" & "$c^j_2 = 0$",

W_9 = "$c^j_2 = 1$",

W_{10} = "$pr_{2,3,\ldots,n+1} x^{\alpha} \neq \langle 0, 0, \ldots, 0\rangle$" & "$pr_1 x^{\alpha} \neq 0$".

The transitions Z_{3+j} $(1 \leq j \leq 4)$ are completely analogous to one another. The j-th of them corresponds to the functioning of the j-th dispatcher. Because the airplanes come in from the j-th direction, or fly out in the j-th direction, the j-th dispatcher certainly gets. Then under his control and at some moment he can be overworked (then $b_j = pr_j x^{\beta} < 0$). That is why when a token moves from place l_{8+j} (the corresponding airplane is waiting for landing under the control of the j-th dispatcher) through the transition Z_{3+j}. After a check of the transition condition, if $b_j \geq 0$ the token moves to place l_{20+j} (the corresponding airplane continues waiting under the control of the same dispatcher), else the token moves to place l_{16+j} and receives as new characteristic the last one but with $c^j_2 = s$, where $b_s = \max_{1 \leq t \leq 4} b_t$ (the corresponding airline will be transferred to the control of a new dispatcher who at the moment control the smallest number of airlines). The transition condition of Z_{3+j} $(1 \leq j \leq 4)$ is

$$r_{3+j} = \begin{array}{c|cc} & l_{16+j} & l_{20+j} \\ \hline l_{8+j} & W_{11} & W_{12} \end{array}$$

where

W_{11} = "$b_j = pr_j x^{\beta} < 0$",

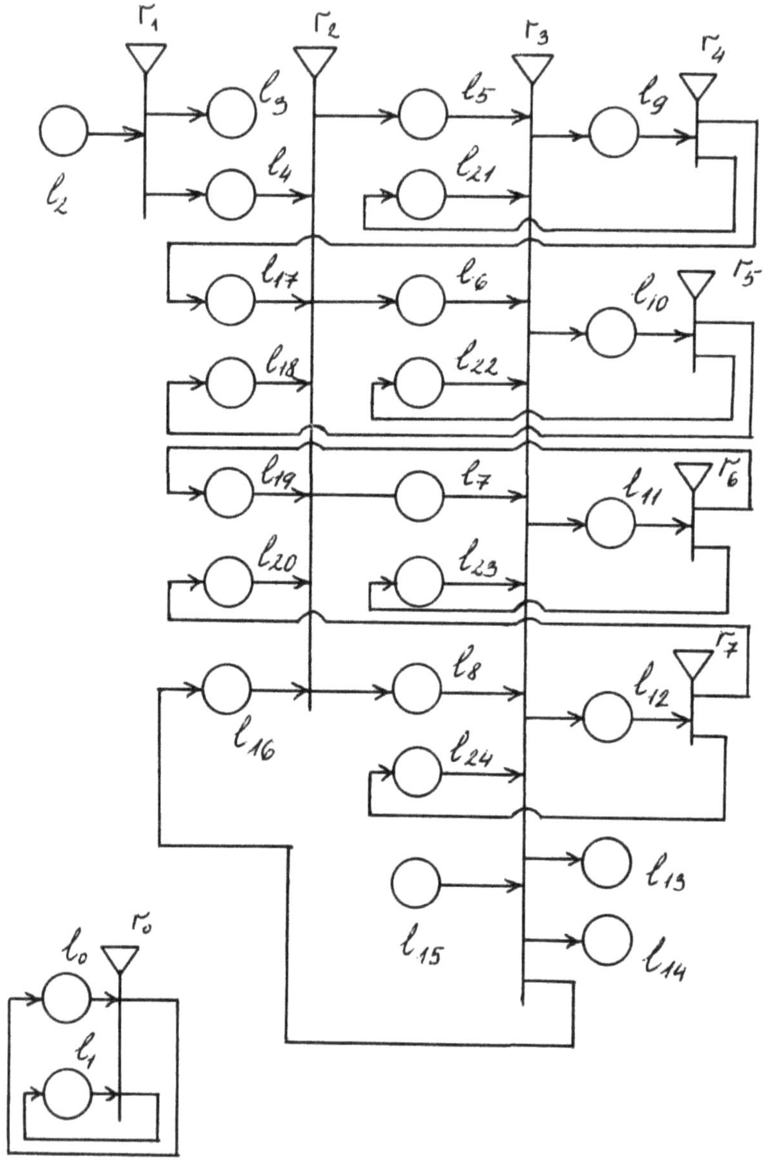

r1
r2
r3
r4
r5
r6
r7
r0
l2
l3
l4
l5
l21
l9
l17
l6
l10
l18
l22
l19
l7
l11
l20
l23
l16
l8
l12
l24
l13
l15
l14
l0
l1

Fig. 3.7

$W_{12} = \neg W_{11}$.

All the transitions and places in the described GN have equal priorities; the tokens in places $l_2, l_3, \ldots, l_{24}$ are ordered as to their second current characteristics in ascending order. The GN starts functioning at a fixed time-moment T and it functions until another fixed moment $T+t^*$. It works tact by tact with an elementary time-step t^o (= 15 sec., e.g.). All transitions are activated at every tact. All arcs have infinite capacities. All transition types are disjunctive.

§ 3.6: GENERALIZED NET MODELS FOR FLEXIBLE MANUFACTURING SYSTEMS

Maria Stefanova-Pavlova and Krassimir Atanassov

§ 3.6.1 Introduction

Flexible Manufacturing Systems (FMSs) [1-58] are an attempt to reconcile the efficiency of the production line with the flexibility of the job shop in order to satisfy a versatile demand at low cost.

Generally two kinds of flexibility are distinguished. The long-term flexibility corresponds to the possibility of introducing a new product group in the manufacturing systems during its operation and with little effort. The short-term flexibility corresponds to the possibility of handling a large variety of product groups at a given time in the manufacturing system [7]. In order to meet these requirements, an FMS is made up of

(a) a set of flexible machines,

(b) an automatic transport system,

(c) a sophisticated decision-making system to decide at each instant what has to be done next and on which machine.

Flexible machines have the capability of performing various operations. They have an automatic tool storage with a retrieval system and the machine programs can be downloaded. An automatic transport system is required to transport the parts for the next operation. The decision making system organizes the production schedules and synchronizes machine utilization. This system is very important for solving problems such as idle time minimization and manufacturing without people, which appear in organizing the work of FMSs.

A FMS is structured into flexible manufacturing cells. Each cell is an elementary system consisting of a flexible machine tool, some local storage facilities for tools and handling devices such as robots in order to transfer parts and tools between the cell and the global transport system [7].

The FMSs are a class of automated manufacturing systems. An FMS comprises processing modules or machines linked by a material handling system, all under central computer control (cf e.g. [3]). The flexibility of such a system is therefore dependent on its components' capabilities and their interconnections as well as on the mode, of operation and control. Some of the basic properties of the FMSs, related to the models below are

(a) the possibility to work autonomously without any human intervention;
(b) automatic realization of basic and supplementary operations;
(c) elimination of production line idle time;
(d) maximum possibility for a complete preparation of a workpiece on a single machine;
(e) high economic efficiency;

and others (cf e.g. [4]).

The rapidly expanding area of FMSs requires new methods for modelling, simulation, control and optimization, as classical methods are not always adequate.

From the mid 80s, (the earliest papers are probably [3-5]) the Petri nets (PNs) which have established themselves as a convenient means for FMS modelling. First of all, the FMS is a discrete system. Any modelling has to be based on the concepts of events and activities. An event corresponds to a state change. When using Petri nets, events are associated with transitions. Activities are associated with the firing of transitions and/or marking of places. Petri nets are appropriate for representing parallelism and synchronisation in an engineering environment. They are very useful at the global coordination level where real-time decision making is required to execute the planned and scheduled production in the manufacturing shop. The real-time decisions of the coordination level also have to be consistent with the manufacturing shop. The Petri net can be seen as a control structure of rule-based system where the rules operate on object attributes (parts, machines, tools) represented by tokens. The Petri net utilization at the global coordination level increases the security and reliability of the FMS Control.

The application of Petri net theory to the FMS is a very rich domain. Let us now briefly sum up some advantages of this approach:

(a) partial order among events can be described to meet flexibility requirements,

(b) states and events are represented explicitly,

(c) a unique family of tools is used throughout the specification, modelling, qualitative validation and performance evaluation, implementation and the operational processes,

(d) the same family of tools is used within various functions of an FMS and at various levels (scheduling, global coordination and local control). It represents an important aid for integrating the whole system,

(e) an accurate, formal description of the synchronization mechanism is provided, which is something essential to achieve security [7],

(f) the graphic nature of Petri nets, their conciseness in comparison with state graphs, the possibility of analysing them , and the existence of techniques for a direct implementation,

(g) the Petri net permits direct realization of the modelled system from the analyzed net [8].

One outstanding aspect of Petri nets is their capacity for being analysed to see if the constructed model is valid. For this the net must satisfy a set of properties which are characteristic of a "good model". Some of the properties are: boundedness - the number of tokens which any place in the net can accumulate during its evolution is bounded; liveness - this does not only mean that the model is deadlock-free, but also that from each state reachable by the net there exist evolutions which can lead to the firing of any transition; reversability - if this is verified it means that it is always possible (from any reachable state) to return the system to its initial state. For analyzing the properties of the Petri nets various techniques have been developed

(a) Reachability graph analysis - this consists of analyzing the net on the basis of the graph of the reachable marking from the initial marking. This method allows a complete analysis of the net.

(b) Reduction analysis - this analysis consists of applying to the net a set of rules which, while preserving the properties that are to be analyzed, gives a smaller model. This method is complementary to the previous method. By first applying this method, later "reachability analysis" is made easier.

(c) Structural analysis - this method allows the analysis of many properties of Petri nets, independently of its initial marking.

The main disadvantage of Petri nets arises from the size of the nets modelling very complex systems, such as FMSs.

Another disadvantage is the fact that a change in the system's functioning forces a substantial change in the net structure.

Another suitable means for an FMS modelling are coloured Petri nets. They consist of places and transitions. A place can be marked by the elements of a subset of colours (none, one or various items of each colour). The tokens of a place will be called coloured tokens or simply, colours. The marking of a place can be represented by a formal sum of colours. The resulting models can be more concise than those obtained from Petri nets and the functions appearing on the net arcs have a clear physical meaning. The main disadvantage of coloured Petri nets is the fact that techniques for model construction, analysis and realization are not yet fully developed. This is a result of this tool's relative newness.

*
* *

3.6.2 *First GN-model of FMSs*

In the PN-models of FMSs constructed so far, the dynamics of the processes in FMSs is described and the data for main time parameters of the modelled processes are investigated (cf. [5]). Analogous results can be obtained by GNs. On the other hand, the constructed generalized net models (GN-models) of FMSs can be used as tools for control of the processes, which function sufficiently slowly in real-time. GN-models also give a possibility of optimization of the controlled process (we shall discuss this problem in a further communication).

Here we shall discuss a process, as described by PNs in [5].

The problem is to develop general understanding of the way in which the performance of an FMS is affected by disturbances and find out how flexibility can attenuate these affects. A GN, modelling the same process is shown on Fig. 3.8. It contains the transitions Z_1, Z_2, Z_3, Z_4, having the following described components. Each of these transitions is activated on every elementary time step (this simplifies the functioning of the GN-model). The capacities of the transition's arcs and of the places are infinite. The transition Z_1 is activated when there is at least one token in the places l_1, l_3, l_{12} and l_{13} and simultaneously there is at least one token in place l_8. The transition Z_2 is activated when there is at least one token in at least one of the places l_2 and l_{14} and simultaneously there is at least one token

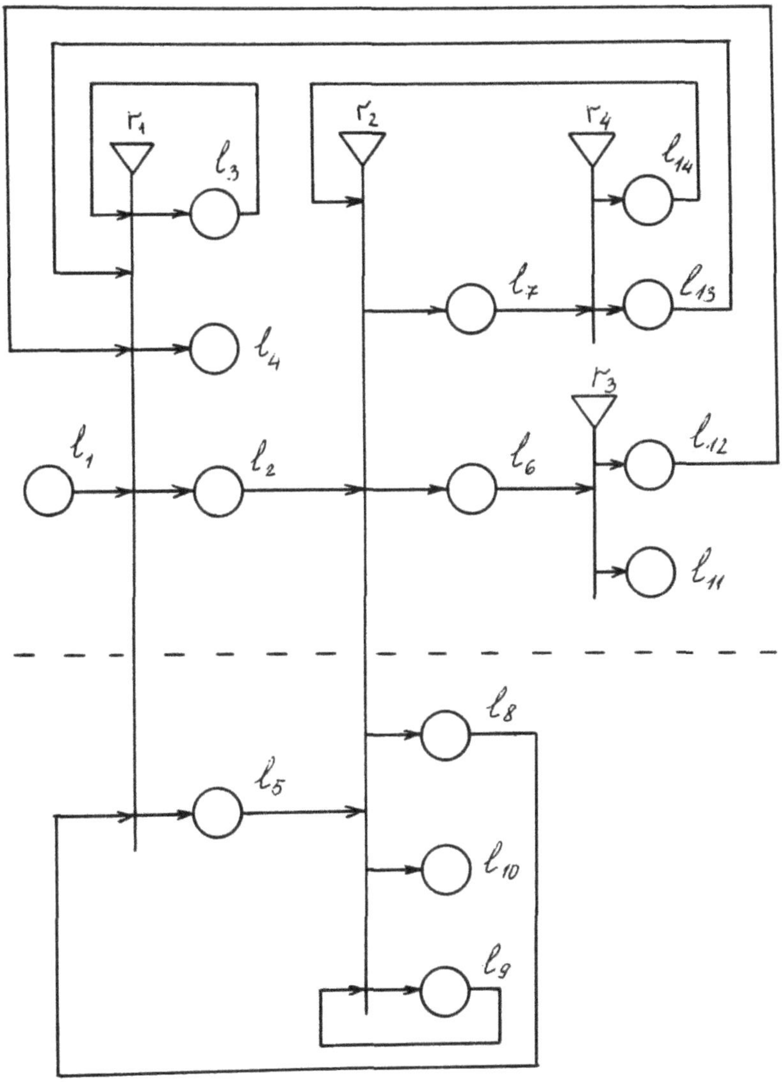

Fig. 3.8

in at least one of the places l_5 and l_9.

The transitions Z_3 and Z_4 are activated when there is at least one token in at least one of their input places. The priorities of the places and transitions are 0 (i.e. they have no priorities). The transition conditions are

$$r_1 = \begin{array}{c|cccc} & l_2 & l_3 & l_4 & l_5 \\ \hline l_1 & W_1 & W_2 & W_3 & false \\ l_3 & W_1 & W_2 & false & false \\ l_{12} & W_1 & W_2 & W_3 & false \\ l_{13} & W_1 & W_2 & W_3 & false \\ l_8 & false & false & false & W_4 \end{array}$$

where

W_1 = "In place l_8 there exists a token with a characteristic which corresponds to the current identifier of an available machine for the token (workpiece)";

$W_2 = \neg W_1$;

W_3 = "In places l_5, l_8 and l_9 there are no tokens with a characteristic which correspond to the current identifier of an available machine for the token (workpiece)";

W_4 = "There exists a workpiece which must be machined";

$r_2 =$

	l_6	l_7	l_8	l_9	l_{10}
l_2	W_5	W_6	false	false	false
l_{14}	W_7	false	false	false	false
l_5	false	false	W_5	W_6	W_8
l_9	false	false	W_5	W_6	W_8

where

W_5 = "The machine which is machining the workpiece is at work";

$W_6 = \neg W_5$;

i. e.

W_6 = "The machine breaks down";

W_7 = "The machine which is machining the workpiece is at work, or there exists another machine (a token in place l_9), which can machine the workpiece";

W_8 = "The machine is shelved";

$r_3 =$

	l_{11}	l_{12}
l_6	W_9	W_{10}

where

W_9 = "The list of the processing machines is exhausted";

$W_{10} = \neg W_9$;

$r_4 =$

	l_{13}	l_{14}
l_7	W_{11}	W_{12}

where

W_{11} = "The repair duration takes a lot of time";

$W_{12} = \neg W_{11}$

In the transfer to the different places, the token assumes the following characteristics (the fact that the characteristic is not given is reflected by the absence from the following list):

Φ_2 -> "an identifier of the available machine ";

Φ_4 -> "the workpiece cannot be machined";

Φ_5 -> "general time for processing of the token in the last token transfer if the token was in place l_8; general time for repair if the token was in place l_9 ";

Φ_6 -> "general idle time, if the token was in place l_{14} ";

Φ_{10} -> "the machine is not able to function";

Φ_{11} -> "general time for processing of the workpiece";

Φ_{12} -> "time for processing with the corresponding available machine".

The GN starts functioning at the time moment T according to a given time-scale and continues t^* elementary time-steps.

At the start of the simulation process in place l_8 there exist tokens with the following characteristics:

"identifier of processing machine; priority of the processing machine; capacity of the processing machine, etc. defined by the GN-model's user".

At different time moments defined by a law (from the user), in the input place l_1 tokens enter with the initial characteristics:

"identifier of the workpiece; identifier's priority; moment of its arrival, a list of processing machines necessary for it; and others defined by the GN-model's user".

The token's transfer in the GN depends on the truth-values of the predicates of the conditions given above. As a result of the transfer, the tokens from both types receive, by the above characteristic functions, the following characteristics: idle time duration in every one of the GN-places, general idle time duration, general time for processing in the GN etc. The tokens of the first type receive as characteristics the values of the general and mean times for the processing of the workpieces from the different types of machines (if these are more than 1), general and mean time for repair, general and mean time for the functioning of the machine between two repairs, etc.

The tokens from the second type receive as characteristics the values of the general and mean times for processing in the different types of machines (if they are more than 1), general and mean waiting time for processing, general and mean time for repair of the processing machine if a breakdown occurs during processing time, etc. The user can define other characteristic function values for the tokens and as a result, corresponding new data can be received. When the real process is controlled by a GN-model in real time, the characteristics of the tokens are obtained from the real environment by the corresponding interface. The truth-values of the predicates of the transition conditions are calculated on the basis of the ones existing in the GN information, accumulated as tokens' characteristics, or as expert's (operator's) subjective estimation, or as a combination of both. On the basis of a concrete GN-model functioning in real time, sufficient information can be accumulated which can be used in the following definitions for changes of values of the random functions. These functions can be applied for GN-simulation of the process. As a result, the GN-simulation processes would be increasingly approximating the real ones and thus the results of the simulation would show the future state of the process. In this case, the initial characteristics of the tokens ought to be taken from the real environment. This information will be necessary at different management levels.

Further improvement of the GN-model of a FMS can be achieved by introducing the influence of the toolroom on its functioning. The condition, flow and management of tools affect in a great extent the performance of the system as far as a researcher wants to investigate more thoroughly the reasons for waiting because of a tool breakage, lack of tools, time delay for tool change, etc.

In order to start machining, a machine tool must have in the tool set all the tools with enough tool-life. Very often waiting for machine tools is due to tool breakages resulting from uncorrectly chosen modes of cutting or from defects in the blank parts. These pauses in the machining of workpieces are of a great significance for the characteristic of the system as a whole.

GN-modelling allows one to introduce all these attributes of the toolroom and tool-dependent conditions of the functioning (i. e. tool-life, tool breakage, availability of tools, time delay due to tool change, etc.) in order to achieve a more realistic picture of the behaviour of the FMS.

*
* *

3.6.3 Second GN-model of FMSs

The constructed model is based on a real functioning system in the plant for metal-cutting machines in Sofia. The system consists of

(a) a store for workpieces, including a station for the main input/output operations where loading/unloading is performed; a fixing zone including fixing stations, intermediate stations (ports), and terminals;

(b) machine park including 12 processing machines of three different types depending on the size of the NC-palets, washing machines for clearing the workpieces, a measuring machine. To eliminate queues in front of the washing machines, only two measure requests are permitted. If a measure request is not strictly determined, the palet with the workpiece waits on the rear buffer of the processing machine while at least one of the washing machines is released.

(c) transport system including robocars in the machine area, robocars in fixing zone, robocars for transportation of tools between the processing machines and the tool store.

A model describing the movement of the workpieces is shown on Fig. 3.9. It consists of four transitions with the following elements:

(a) transition Z_1 with input places l_1, l_2 and l_8 and type $v(l_1, l_2, l_8)$ and output places l_2, l_3 and l_4;

(b) transition Z_2 with input places l_3 and l_{12} and type $v(l_3, l_{12})$ and output places l_5, l_6, l_7, l_8.

(c) transition Z_3 with input places l_5, l_6, l_7 and l_9 and type $v(l_5, l_6, l_7, l_9)$ and output places l_9, l_{10}.

(d) transition Z_4 with input place l_{10} and output places l_{11}, l_{12}.

The capacities of the transition's arcs are 1, and those of the places are ∞ with the exception of places l_5, l_6, l_7 (which have capacities equal to the numbers of the processing machines for small, middle and big workpieces) and l_{10} with a capacity equal to 2 (there are no more than two measure requests).

Every transition is activated on every elementary time-step.

In this model the transitions are activated when at least one

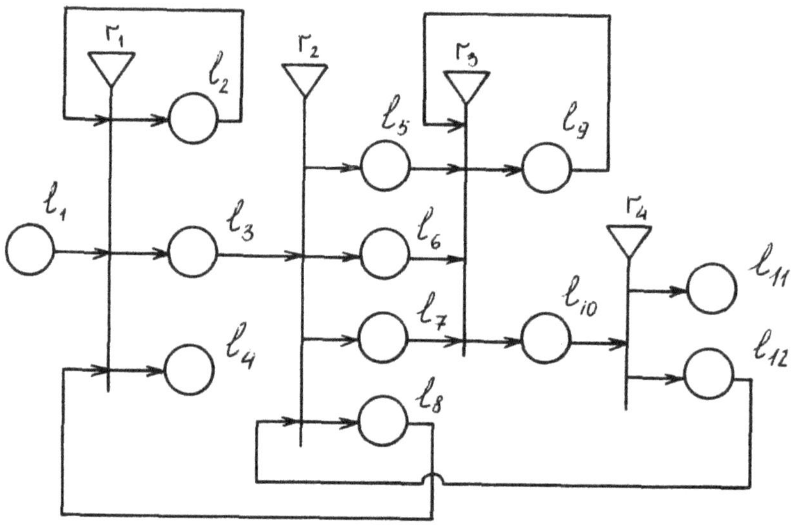

Fig. 3.9

of their input places contains at least one token.

The transition conditions are

$r_1 =$

	l_2	l_3	l_4
l_1	W_1	W_2	false
l_2	W_1	W_2	W_3
l_8	W_1	W_2	W_3

where

W_2 = "there exists a machine which can process the workpiece";

$W_1 = \neg W_2$;

W_3 = "the list of the processing machines is exhausted";

$$r_2 = \begin{array}{c|cccc} & l_5 & l_6 & l_7 & l_8 \\ \hline l_3 & W_4 & W_5 & W_6 & false \\ l_{12} & W_4 & W_5 & W_6 & W_7 \end{array}$$

where

W_4 = W′ v (¬W′ & W″) v (¬W′ & ¬W″ & W_6);

W_5 = W″ v (¬W″ & W_6);

W_6 = "the workpiece is big";

W_7 = "the workpiece is going for a further processing or to the store";

and

W′ = "the workpiece is small";

W″ = "the workpiece is a medium in size";

$$r_3 = \begin{array}{c|cc} & l_9 & l_{10} \\ \hline l_5 & W_8 & W_9 \\ l_6 & W_8 & W_9 \\ l_7 & W_8 & W_9 \\ l_9 & W_8 & W_9 \end{array}$$

where

W_8 = "the measuring machine is not free";

W_9 = ¬W_8;

$$r_4 = \begin{array}{c|cc} & l_{11} & l_{12} \\ \hline l_{10} & W_{10} & W_{11} \end{array}$$

where

W_{10} = "the workpiece is shelved";

W_{11} = $\neg W_{10}$.

During their movement through the net, the tokens obtain new characteristics by the characteristic function as follows:

Φ_3 -> "waiting-time for place l_2",

Φ_4 -> "general time for processing of the workpiece", "difference between the real time and the time which is given in the initial characteristic",

Φ_5 -> "name of the processing machine for small workpieces",

Φ_6 -> "name of the processing machine for medium-size workpieces",

Φ_7 -> "name of the processing machine for big workpieces",

Φ_9 -> "time for processing in 15 or 16 or 17, respectively",

Φ_{10} -> "waiting time in l_9",

Φ_{11} -> "value of the consuming labour",

Φ_{12} ->"time for processing separate steps".

The places represent the following process components:

l_1 -> initial status of the process,

l_2 -> a workpiece is waiting in the store for processing,

l_3 -> the workpiece is going for processing,

l_4 -> the workpiece is going out of the system after processing,

l_5 -> the workpiece is processed on a machine for small workpieces,

l_6 -> the workpiece is processed on a machine for medium-size workpieces,

l_7 -> the workpiece is processed on a machine for big workpieces,

l_8 -> the workpiece is going back to the store,

l_9 -> the workpiece is waiting for releasing of the washing machines,

l_{10} -> the workpiece is washed and measured from the measuring machine,

l_{11} -> the size of the workpiece does not meet the requirements, the workpiece is shelved and going out of the system,

l_{12} -> the workpiece is going for further processing, if necessary it is going to the store.

* * *

§ 3.6.4. Third GN-model of FMSs

The system to be controlled is part of a flexible workshop of the Renault company. The purpose of the system is to mastic seal the body work of automobiles. For this purpose it has six workstations (or simply stations), each consisting of one work place P, sufficient for one car body, and also one transfer bench with two roller tables, one designed to load the station (TL) and the other to unload it (TU). The bench, which is mobile, can adopt two positions: left, to load, or right to unload the car bodies.

A system of transport, made up of six roller tables (RT_1, ..., RT_6) has the task of transporting all car bodies which enter the system (via table RT_1) to the sealing station, and of later unloading and removing them to the exit via RT6. Each table (TL, TU, RT_i) can carry only one car body, and at any station there can be a maximum of two (P-TU, P-TL, or TU-TL) at a given time [27].

A coloured Petri net model of this system is applied. In modelling the autonomous coloured Petri net model, we can distinguish three stages:

(a) independent modelling of different subsystems, which form part of the system under consideration, using a coloured Petri net. In the above-mentioned system, we can distinguish two subsystems: transport and stations;

(b) fusion of the modules by fusing the transitions which represent synchronization between the subsystems. In the system, the loading and unloading of the stations requires synchronization between the transport and the station subsystems;

(c) the eventual simplification of the coloured Petri net, obtained by eliminating implicit places. These are places whose marking can be expressed as a positive linear combination of a set of places of the net and whose elimination preserves the liveness and boundedness of the net.

With each place, we associate a set of colours such that we can distinguish all the tasks and resources associated with it. The colours can be simple or compound (n-tuples) depending on whether the associated tasks or resources are defined by one or several parameter With each transition, we associate a set of colours such that its cardinal is equal to the number of different possible evolutions of the state of the system which are defined by their possible firings [27].

Below we shall construct a GN which describes the above process modelled by a coloured Petri net.

The graphical structure of this net is given on Fig 3.10.

The GN has no local temporal components. Every transition is activated when there is at least one token in its input places, i.e. all transition types are of a disjunctive type.

The capacities of the places are (cf. [27]):

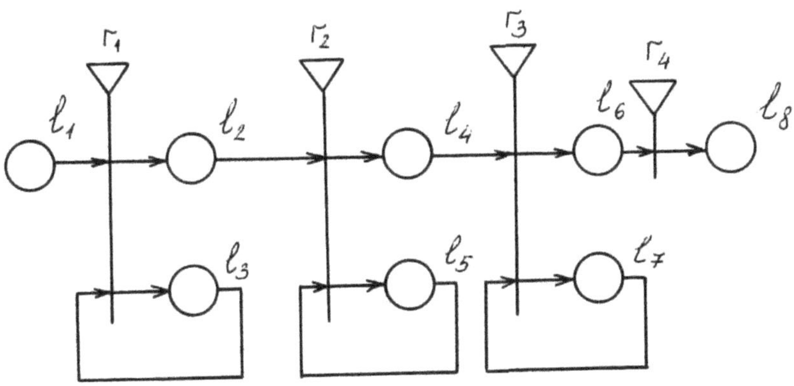

Fig. 3.10

$$c(l_1) = c(l_8) = \infty, \quad c(l_2) = \ldots = c(l_7) = 6.$$

The priorities of the places are

$$\pi_L(l_3) = \pi_L(l_5) = \pi_L(l_7) >,$$

$$\pi_L(l_1) = \pi_L(l_2) = \pi_L(l_4) = \pi_L(l_6) = \pi_L(l_8).$$

The priorities of the transitions are equal.

The capacities of the arcs are "∞".

Every token enters the net at a certain time-moment (determined by a function θ_K) according to some fixed time-scale (with initial time-moment T, elementary time-step t^o and time-duration t^*). It can receive an undetermined number of characteristics and enters with an initial characteristic "type of the car body, current number (and others)".

The GN contains 4 transitions and 8 places. They contain the following transition conditions and to the GN places, the following characteristic functions are associated.

$$r_1 = \begin{array}{c|cc} & l_2 & l_3 \\ \hline l_1 & W_1 & W_2 \\ l_3 & W_1 & W_2 \end{array}$$

W_1 = "there is an empty loaded table" & "there is an empty roller table";

$W_2 = \neg W_1$;

Φ_2 -> "<number of the empty loaded table, duration of the transfer from the input place of the modelled system to the loaded table, duration of waiting in place l_3>";

Φ_3 -> *,

$$r_2 = \begin{array}{c|cc} & l_4 & l_5 \\ \hline l_2 & W_3 & W_4 \\ l_5 & W_3 & W_4 \end{array}$$

W_3 = "the workplace with a number $pr_1 x^{\alpha}_{cu}$ is empty";

$W_4 = \neg W_3$;

Φ_4 -> "<duration of the transfer from the loaded table to the workplace, duration of waiting in place l_5>";

Φ_5 -> *;

$$r_3 = \begin{array}{c|cc} & l_6 & l_7 \\ \hline l_4 & W_5 & W_6 \\ l_7 & W_6 & W_7 \end{array}$$

W_5 = "the unlaoded table with a number $pr_1 x^{\alpha}_{cu-1}$ is empty",

$W_6 = \neg W_5$,

Φ_6 -> "<duration of the sealing of the car body, duration of waiting in place l_7, duration of the transfer from the workplace to the unloaded table">

Φ_7 -> *,

$$r_4 = \begin{array}{c|c} & l_8 \\ \hline l_6 & W_7 \end{array}$$

where

W_7 = "there is an empty roller table"

Φ_8 -> "<duration of the transfer from unloaded table to roller table">.

*
* *

§ 3.6.5 GN-model of flexible manufacturing cell

This model concerns the control system organisation of a flexible manufacturing cell. We shall show how the Petri net model of the flexible manufacturing cell from [28] can be represented by GNs.

A flexible manufacturing cell is used for automatic production of metal parts with specific dimensions. It is characterised by an increasing automation of technological operations, a better organization and a more accurate control finalization strategy aimed at increasing production. The configuration of a flexible manufacturing cell, whose functioning is controlled by this model, contains: robots R_1 and R_2, machine tools (NC_1-milling machine, NC_2-lathe, NC_3-press), two auxiliary buffers (B_1 and B_2), incoming T_1 and outgoing T_2 transporters, a bin for rejected parts.

The cell works in the following way: Raw pieces of material come by the incoming transporter T_1 up to the take up position, which is defined by a step movement. Robot R_1 waits in front of the transporter T_1, grips the object and goes to the working space of a milling machine NC_1 and there leaves the workpiece. After that, robot R_1 draws back and waits for a milling machine to complete the processing. After processing the object, the robot goes to the milling machine, takes the processed object and carries it over to a free position in buffer B_1. Robot R_2 takes the workpiece from the buffer B_1 and carries it over to the lathe. Then robot R_2 draws back. When the processing on lathe NC_2 is completed, R_2 takes the piece from the lathe and transfers it to the press NC_3. Then, the robot R_2 withdraws in front of the press till the workpiece is pressed. After that the robot R_2 takes the workpiece and carries it over to the transporter T_2. Buffer B_2 is used for temporary laying down of the workpieces which the lathe has finished and waits to be pressed. The reject store is used for storing the workpieces ruined during machining.

A GN-model, describing this process is shown in Fig. 3.11. The transition conditions in the form of matrix of predicates are

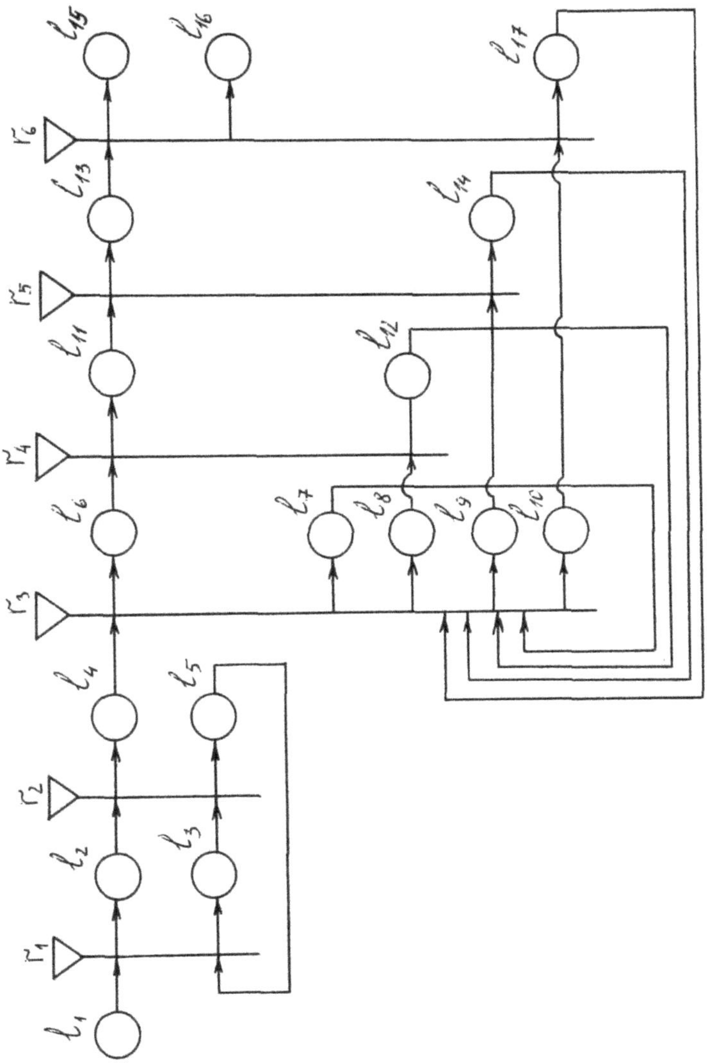
l1
r1
l2
l3
r2
l4
l5
r3
l6
l7
l8
l9
l10
r4
l11
l12
r5
l13
l14
r6
l15
l16
l17

Fig. 3.11.

$$
r_1 = \begin{array}{c|cc}
 & l_2 & l_3 \\ \hline
l_1 & W_1 & false \\
l_2 & false & W_1
\end{array}
$$

where

W_1 = "$\partial(l_2$, TIME) = $\partial(l_2)$"

(the function $\partial(l)$ gives the capacity of the place l and the function $\partial(l$, TIME) gives the number of tokens in the place l);

$$
r_2 = \begin{array}{c|cc}
 & l_4 & l_5 \\ \hline
l_2 & W_2 & F \\
l_3 & F & W_2
\end{array}
$$

where

W_2 = "$\partial(l_2$, TIME) < $\partial(l_2)$" & "the necessary time for processing in place l_2 is run out".

$$
r_3 = \begin{array}{c|ccccc}
 & l_6 & l_7 & l_8 & l_9 & l_{10} \\ \hline
l_4 & W_3 & false & false & false & false \\
l_7 & false & W_3 & \neg W_3 \& W_4 & \neg W_3 \& \neg W_4 \& W_5 & \neg W_3 \& \neg W_4 \& \neg W_5 \& (W_6 \vee W_7) \\
l_{12} & false & W_3 & \neg W_3 \& W_4 & \neg W_3 \& \neg W_4 \& W_5 & \neg W_3 \& \neg W_4 \& \neg W_5 \& (W_6 \vee W_7) \\
l_{14} & false & W_3 & \neg W_3 \& W_4 & \neg W_3 \& \neg W_4 \& W_5 & \neg W_3 \& \neg W_4 \& \neg W_5 \& (W_6 \vee W_7) \\
l_{17} & false & W_3 & \neg W_3 \& W_4 & \neg W_3 \& \neg W_4 \& W_5 & \neg W_3 \& \neg W_4 \& \neg W_5 \& (W_6 \vee W_7)
\end{array}
$$

where

W_3 = "$\partial(l_6$, TIME) < $\partial(l_6)$",

W_4 = "$\partial(l_{11}, TIME) < \partial(l_{11})$" & "the necessary time for processing in place l_6 is run out",

W_5 = "$\partial(l_{13}, TIME) < \partial(l_{13})$",

W_6 = "the workpiece is good",

W_7 = "the workpiece is ruined".

$$r_4 = \begin{array}{c|cc} & l_{11} & l_{12} \\ \hline l_6 & W_4 & false \\ l_8 & false & W_4 \end{array}$$

$$r_5 = \begin{array}{c|cc} & l_{13} & l_{14} \\ \hline l_{11} & W_5 & false \\ l_9 & false & W_5 \end{array}$$

$$r_6 = \begin{array}{c|ccc} & l_{15} & l_{16} & l_{17} \\ \hline l_{13} & W_6 & W_7 & false \\ l_{10} & false & false & W_7 \end{array}$$

In conclusion we shall note that the above constructed GN-model has important advantages in relation to the other Petri net models

(a) it is considerably more universal;

(b) as a result of simulation of a flexible manufacturing cell with this model, more information on the process can be received (the tokens which enter the net with certain initial characteristics obtain new characteristics while moving through the net by the characterizing functions associated with the places);

(c) the process (if it proceeds sufficiently slowly) can be controlled on the basis of the constructed model;

(d) the process can be optimized on the basis of the model.

*
* *

3.6.6 Queue models and GN-models in flexible manufacturing

Scheduling is one of the important problems in the functioning of the FMSs.

The classical scheduling literature dating back to the 1960s, has formulated tactical scheduling as the following optimization problem: a job shop has n jobs and m machines. Each job has a routing composed of a sequence of m operations, a unique machine being assigned to each. *The goal is to find a schedule (the start time of mn operations)*, which minimizes some criteria, e.g. makespan or tardiness. A simple approximation of the computational complexity can be calculated as follows: a job has m operations, one on each machine, and since each machine can be scheduled in n! ways, it follows that there are $(n!)^m$ schedules. A method for finding the optimal schedule would be the enumeration of all these schedules and choosing the best one. The search space, consisting of the possible number of valid schedules is very large. Traditionally, scheduling has been considered the last phase of the planning process, which gives a solution of the problem as when to allocate a particular resource to an operation [59].

In practice this problem is solved in the following way:

(a) the priority of the job is defined;

(b) the jobs are ordered according to their priorities;

(c) if the planned resource is not available, the corresponding jobs with a lower priority must be transferred.

In flexible manufacturing the machines are ordered in such a way that they can meet the requirements of full processing of a workpiece in one place.

The FMSs can be considered as one capacity unit. Because of centralizing according to objects, but not according to functions, the material flow can be easily observed. In this way the transfer of a job does not attach to the other capacity units.

To solve the scheduling problems in the FMSs, some methods are used:

(a) queue models;

(b) methods of mathematical programming;

(c) methods with priority rules.

The GNs are also used for modelling the functioning of the FMSs.

We shall try to compare the queue models with the GN-models according to the method of solving scheduling problems in FMSs.

The queue model consists of knots (machines) with a queue of workpieces in front of them. The time for processing is random. The workpieces are processed according to the order of their entrance into the system (first in, first out) or in a random way. After processing, the workpiece goes to another machine with some probability. Analytical results are as follows: middle grade of loading up the machines, middle quantity of workpieces in the system, middle quantity of waiting workpieces [60].

One of the main disadvantages of the queue models for modelling FMSs is the value of the analytical results. These results do not match the real situation.

It is due to several factors:

(a) the often-used exponential distribution for the processing time is not close to the real distribution; it is possible to take into account deterministic processing time-moments only when the processing times are identical;

(b) the processing sequence is strictly determined;

(c) for the queue models an unlimited capacity is implied; this fact does not match with the capacities of the machine buffer and storage;

(d) only one priority rule for scheduling is used (FIFO - first in, first out);

(e) the workpieces cannot be individually traced, the description being for the whole system;

(f) the disturbances (machine breakdown or lack of tools) cannot be taken in account.

The GN-model of FMSs removes the drawbacks of queue models:

(a) we can use an arbitrary distribution of the processing times so that in future extensions of the GNs one distribution can be exchanged with another;

(b) the processing sequence is not strictly determined; it depends on the optimal choice at a particular time-moment;

(c) the capacities correspond to the real system (machine buffer, storage);

(d) different priority rules can be used - rules for the optimal choice of the workpieces;

(e) the workpieces can be individually traced;

(f) the disturbances can be taken into account in the model.

Comparing the two approaches of modelling FMSs, we can conclude that the GN-models are more suitable for deciding the sheduling problems in flexible manufacturing.

*
* *

3.6.7: Conclusion

In conclusion, note that the above constructed GN-models have the following important advantages in relation to other PN-models.

i) They are considerably more universal compared with other PN-models. A PN-model is constructed for a concrete type of FMSs. The GN-model discussed here describes the functioning of the results of the work of every element of a complete class of FMSs.

2) As a result of the simulation of the FMS with this GN-model more information about the process can be extracted. Note that the characteristic functions defined are constructed in such a way that they are maximally similar to those for the PN-model.

3) The process (if it proceeds sufficiently slowly) can be controlled on the basis of the constructed model.

4) The process of the transfer of the workpieces and the choice of the processing machines can be optimized on the basis of the GN-model.

This paper is based on [61-64].

REFERENSES:

[1] J. Buzacott and J. Shanthikumar, Models for understanding flexible manufacturing systems. AIIE Transactions, 1980, 12, 339 p.

[2] V. Vassilev, An organization, control and economy of flexible manufacturing production in manufacturing, Moscow, "Machinostroenie", 1986, 312 p. (in Russian).

[3] D. Dubois and K. Stecke, Using Petri nets to represent production process, Proc. of IEEE Conf. on Decision and Control, Dec. 1983, 1062.

[4] P. Alanche, F. Benzakour, F. Dolle, P. Gollet, P. Rodrigues and R. Vallette, PSI: A Petri net based simulator for flexible manufacturing systems, Advances in Petri nets (ed. by G. Rozenberg et al.) in Lectures Notes in Computer Science, Vol. 188, 1985, 1-14.

[5] M. Barad and D. Sipper, Flexibility in manufacturing systems: definitios and Petri net modelling, Int. J. Prod. Res., Vol. 26, 1988, No. 2, 237-248.

[6] H.-J. Foerster, Entwicklung von Anforderungsprofilen flexibel automatisierter Fertigungskonzepte an die Produktionsplanung und -Steurung, Schlussbericht, Aachen 1987.

[7] M. Silva and R. Valette, Petri nets and flexible manufacturing, Advances in Petri nets 1989 (G. Rozenberg, Ed.), Lecture Notes in Computer Science 424, 1990, 374-417.

[8] J. Martinez, H. Alla, M. Silva, Petri nets for the specification of FMSs, Modelling and design of Flexible Manufacturing Systems, (A. Kusiak, Ed.), Elsevier Science Publishers B.V., Amsterdam, 1986, 389-406.

[9] J. Ahuja mand K. Valavanis, A hierarchial modeling methodology for flexible manufacturing systems using extended Petri nets. 1988 International Conference on Computer Integrated Manufacturing - Washington, DC, USA: IEEE Comput. Soc. Press, 1988, 350-356.

[10] A. Bobbio and A. Savant, Modelling automated production systems by deterministic Petri nets. IFS Simulation in Manufacturing, 3-rd Intern. Conf., 1987, 127-136.

[11] H. Boerner and I. Pritika, Aspekte der Modellierung von Flexiblen Fertigungssystemen (FFS) auf Basis von Petri-Netzen. Wissenschaftliche Zeitschrift der Technischen Universitaet Karl-Marx-Stad, Vol. 29. No. 6, 1987, 793-797.

[12] F. Bukanov and K. Belikov, Technology of design flexible manufacturing Systems by means of coloured Petri Nets. Systems of Robot and Flexible Automation, Kuibishev Politechnical Institute, 1987, 62-65 (in Russian).

[13] F. Capkovic, Decision support algorithm for flexible manufacturing Systems Control. Computers in Industry, Vol. 10, No3, 1988, 165-170.

[14] H. Carstensen, Decidability Questions for Fairness in Petri Nets. Lecture Notes in Computer Science, Vol. 247, STACS 87, 4th Annual Symposium on Theoretical Aspects of Computer Science, Passau,

1987, Proceedings (F. J. Brandenburg, G. Vidal-Naquet and M. Wirsing, Eds.), Springer Verlag 1987, 396-407.

[15] B. Descotes-Gennon and F. Hemon, Mercier des Rochettes, R. ISATIS: An Editor-Simulator of Manufacturing Systems Using Timed coloured Petri Nets. Databases for Production Management. Proceedings of the IFIP TC5/WG / Working Conference, 1989, Barcelona, Spain - ENSIEG, St. Martin d'Heres, F. ASTEC, Grenoble, 1989, 333-348.

[16] K. Srihari, C. Emerson and J. Cecil, Modeling Manufacturing with Petri Nets. CIM Review, Vol. 6, No. 3, 1990, 15-21.

[17] S. Teng and J. Black, Cellular Manufacturing Systems Modeling: the Petri Net Approach. Journal of Manufacturing Systems, Vol. 9, No. 1, 1990, 45-54.

[18] S. Tzafestas, Petri-Net and Knowledge-Based Methodologies in Manufacturing Systems Modelling Simulation and Control. Computer Integrated Manufacturing. Proceedings of the 5th CIM Europe Conference, 1989, Athens, Greece (G. Halatsis, et al., Eds.), Kempston, UK, IFS Publications, 1989, 39-50.

[19] K. Jensen, Coloured Petri Nets. Lecture Notes in Computer Science Vol. 254: Petri Nets: Central Models and Their Properties, Advances in Petri Nets, 1986, Part I, Proceedings of an Advanced Cource, Bad Honnef, 1986 (W. Brauer, W. Reisig and G. Rozenberg, Eds.), Springer-Verlag, 1987, 248-299.

[20] F. Itter, Integrierte Modellbildung und Simulation mit Petrinetzen. Zeitschrift for wirtschaftliche Fertigung und Automatisierung ZWF/CIM, Vol. 84, No. 2, 1989, 90-92.

[21] F. Janzen, H. Kath , M. Mohrle and H. Seifert, Petrinetze in der Produktionstechnik. Zeitschrift for wirtschaftliche Fertigung und Automatisierung ZWF/CIM, Vol. 84, No. 3, 1989, 141-145.

[22] S. Menon and P. Ferreira, Coloured Petri net based architecture for coordination control of flexible manufacturing systems. Proc. of "Advances in Manufacturing System Engeneering" presented at

the Winter Annual Meeting of the ASME, 1988, Chicago, IL, USA - New York, NY, USA: ASME, 1988, 69-88.

[23] S. Menon and P. Ferreira, Analysis of coloured Petri net based models for coordination control of flexible manufacturing systems. Proceedings of SME 17th NAMRC, Ohio State Univ., May 1989, 331-338.

[24] A. Mowafak Hassan, The modelling and analysis of flexible manufacturing systems using Petri nets. Ph.D. Thesis, Technical University of Wroclaw, Institute of Industrial Engeneering and Management (Oct., 1988).

[25] Y. Nakamura, I. Hatono, Y. Kohara, K. Yamagat and H. Tamura, FMS scheduling using timed Petri net and rule base. Proceedings of the USA-Japan Symposium on Flexible Automation - Crossing Bridges: Advances in Flexible Automation and Robotics, 1988, Minneapolis, MN, USA, 883-890.

[26] U. Negretto, A functional process model for flexible assembly cells. Proceedings of the 19th International Symposium on Automotive Technology and Automation (ISATA), 1988, Monte Carlo, Monaco; with Particular Reference to Cell Control and Quality Management Systems for the Manufacturing Industries, Vol. 1. - Croydon, UK. Allied Autom., 427-443.

[27] J. Martinez and M. Silva, A language for the description of concurrent systems modeled by colored Petri nets: Application to the control of flexible manufacturing systems, IEEE Workshop on Languages for automation, New Orleans, Nov. 1984, 72-77.

[28] M. Jocovic, An application of Petri-nets in the controlsystem of the FTC, Microprocessing and Microprogramming 24 (1988), 681-686.

[29] J. Campos, J. Colom and M. Silva, Performance evaluation of repetitive automated manufacturing systems, Proc. of Second International Conference on Computer Integrated Manufacturing, New York, May 1990, 74-81.

[30] P. Muro, J. Villarroel and M. Silva, A knowledge representation environment for manufacturing control systems design and prototyping, 6th Symposium on information control problems in manufacturing technology (E. Puente, Ed.), Madrid, September 1989, 585-589.

[31] S. Velilla and M. Silva, The spy: a mechanism for safe implementation of highly concurrent systems, Proceedings of the 15-th IFAC/IFIP Workshop "Real Time Programming'1988" (A. Crespo and J. A. de la Puente, Eds.), Valencia, Spain, May 1988, 75-81.

[32] J. Villarroel, J. Martinez and M. Silva, GRAMAN: A Graphic system for manufacturing system design, Proceedings of the IMACS Symposium on System Modellingand Simulation (S. Tzafestas, A. Eisenberg and L. Carotenuto, Eds.), Cetraro, Italy, September, 1988, 311-316.

[33] P. Muro and J. Villarroel, KRON:Redes orientadas a la representacion del concomiento, 3 Reunion Tecnica de la Asociacion Espanola para la Inteligencia Artificial, ACTAS, Madrid de noviembro de 1989, 3-22.

[33] U. Negretto and M. Rillo, Erweiterte Petri-Netze in flexiben Fertigungs- and Montagesystemen, Robotersysteme, Vol. 4, No. 1, 1988, 34-42.

[34] Y. Narahari and N. Viswanadham, A Petri net approach to the modelling and analysis of flexible manufacturing systems, Annals of Operations Research 3, 1985, 449-472.

[35] H. Alayan and R.-W. Newcomb, Petri-net robot models for robot networks, IEEE, Vol. 3, 1986, 996-999.

[36] F. Archetti, Performance evaluation of flexible manufacturing systems using Petri nets. One-Day Seminar at the Bocconi University of Milan "Applicability of Petri nets to Operations Research", 1986, 143.

[37] C. Beck and B. Krogh, Models for simulation and discrete control

of manufacturing systems, Proceedings 1986 IEEE International Conference on Robotics and Automation, San Francisko, 1986, IEEE Comput. Soc. Press, Washington, 1986, 305-310.

[38] J. Martinez and M. Solva, New methods for specifying controls of FMS, Autominstrum (Spain), Vol. 20, No. 156, 185-195 (in Spain).

[39] M. Kamath and N. Viswanadham, Applications of Petri nets based models in the modelling and analysis of flexible manufacturing systems, Proceedings 1986 IEEE International Conference on Robotics and Automation, San Francisko, 1986, IEEE Comput. Soc. Press, Washington, 1986, 312-317.

[40] R. Ravichandran, A. Chakravarty, Design support in flexible manufacturing systems using timed Petri nets, Journal of Manufacturing Systems, Vol. 5, No. 2, 89-101.

[41] M. Shaw and A. Whinston, Task bidding and distributed planning in flexible manufacturing, The Second Conf. on AI Applications, 1985, 184-189.

[42] M. Silva and J. Martinez, A software environment for designing with colored Petri nets and their implementation. Application to flexible manufacturing systems. One-Day Seminar at the Bocconi University of Milan "Applicability of Petri nets to Operations Research", 1986, 95-127.

[43] T. Klotz, R. Phillips, R. Sprattling and H. Sutherland, Real-time performance evaluation of local-area networks used in automated manufacturing systems, Proceedings 1986 IEEE International Conference on Robotics and Automation, San Francisko, 1986, IEEE Comput. Soc. Press, Washington, 1986, 1723-1730.

[43] R. Valette, M. Courvoisier, H. Demmou, J. Bigou and C. Desclaux, Putting Petri nets to work for controlling flexible manufacturing systems. Proceedings of 1985 IEEE International Symposium on Circuits and Systems, New York, 1985, 929-932.

[44] J. Bourney and J. Gentina, Structuring of the procedural part of

the control system of flexible manufacturing cells, Congres Automatique 1988, Grenoble, 233-242 (in French).

[45] P. Stotts, R. Newcomb and Z. Ning Cai, Modelling the logical structure of flexible manufacturing systems with Petri nets, Computer Communications, Vol. 12, No. 4, 1989, 193-203.

[46] H. Tamura, K. Yamagata and I. Hatono, Decision making for flexible manufacturing-OR and/or AI approaches in scheduling. Systems Analysis, Modelling, Simulation, Vol. 6, No. 5, 1989, 363-371.

[47] J. Ahuja and K. Valavanis, Modified Petri nets and comprehensive modeling of flexible manufacturing systems. Proc. of the 12-th IMACS World Conf., Vol. 3, 1988, 532-534.

[48] J. Barbez, E. Craye, J. Gentina and J. Mayet, Hierarchical level and implementation for analysis and synthesis of control and reliability of flexible manufacturing systems. Proc. of the 12-th IMACS World Conf., Vol. 3, 1988, 552-558.

[49] J. Gentina and D. Corbeel, Petri nets in manufactoring, System & Control Encyclopedia: Theory, Technology, Applications, Vol. 6 (M. Singh, Ed.), Pergamon Press, Oxford,, 1987, 3670-1673.

[50] R. Mercier Des Rochette, B. Descotes-Genon and P. Ladet, Modelling and F.M.S. Control implementation, Proc. of the 12-th IMACS World Conf., Vol. 3, 1988, 559-661.

[51] A. Pagnoni, Computer management of coordination in flexible manufacturing systems. Proc. of the 12-th IMACS World Conf., Vol. 3, 1988, 500-502.

[52] W. Zhang, Automatic robot planning by marked Petri nets. Proc. of the Int. Workshop on AI for Industrial Applications, New York, 1988, 358-361.

[53] H. Kodate, K. Fujii and K. Yamanoi, Representation of FMS with Petri net graph and its applications of simulation of system operation. Robotics & Comput. Integrated Manufacturing, Vol. 3, No. 3, 1987, 275-283.

[54] G. Balbo, G. Chiola, G. Franceschinis and G. Roet, Generalized stochastic Petri nets for the performance evaluation of FMS, Proceedings 1987 IEEE International Conference on Robotics and Automation, IEEE Comput. Soc. Press, Washington, 1987, 1013-1018.

[55] A. Likic and V. Zivkovic, A software package for representation and analysis of flexible manufacturing systems based on Petri nets, Proc. of the 3rd European Simulation Congress, 1989, Edinburg, (D. Murray-Smith, Ed.), 502-507.

[56] M. Silva, Petri nets and flexible manufacturing, Lecture Notes in Computer Science, Vol. 424, 1990, 374-417.

[57] A. Zebiri, M. Bonney and M. Head, The use of Petri nets as a basis for simulating manufacturing systems, Proc. of the 3rd European Simulation Congress, 1989, Edinburg, (D. Murray-Smith, Ed.), 825-829.

[58] F. Archetti and A. Sciomachen, Representation, analysis and simulation of manufacturing systems by Petri net based models, Discrete Event Systems: Models and Applications. IIASA Conf., (P. Varaiya and A. Kurzhanski, Eds.), Springer-Verlag, 1988, 162-178.

[59] R. Stokey, AI Factory Scheduling: Multiple Problem Formulation, Sigart Newsletter, October 1989, Number 110, 27-29.

[60] G.-W. Hintz, Ein wissensbasiertes System zur Produktionsplannung und -steurung fuer flexible Fertigungssysteme, VDI-Verlag, Duesseldorf, 1987.

[61] K. Atanassov, E. Dincheva, D. Matev, M. Stefanova-Pavlova, Generalized net-representation of the flexible manufacturing systems, Proc. of the 14 Symp. on Operation Research, Ulm, Sept. 1989, 521-528.

[62] M. Stefanova-Pavlova, Generalized net representation of manufacturing system, First Sci. Session of the Math. Found. AI Seminar, Sofia, Oct. 10, 1989, Preprint IM-MFAIS-7-89, Sofia,

1989, 60-62.

[63] M. Stefanova-Pavlova, R. Christov, E. Dincheva, Generalized net net model of a flexible manufacturing system, Second Sci. Session of the Math. Found. AI Seminar, Sofia, March 30, 1990, Preprint IM-MFAIS-1-90, Sofia, 1990, 40-42.

[64] M. Stefanova-Pavlova, Queue models and Generalized net models in flexible manufacturing. Third Sci. Session of the Math. Found. AI Seminar, Sofia, June 12, 1990, Preprint IM-MFAIS-2-90, Sofia, 1990, 13-14.

§ 3.7: GENERALIZED NET MODELS OF THE ACTIVITY OF NEFTOCHIM PETROCHEMICAL COMBINE IN BOURGAS

Stela Dimitrova, Ludmila Dimitrova, Trajana Kolarova,
Plamen Petkov, Krassimir Atanassov and Rumen Christov

Generalized Nets (GNs) are a comfortable means for designing complex objects which are characterized by a large scale of various parallel processes taking place in the real time. In the general case, it is impossible to show analytical formulas formalizing these processes. For their modelling, various simulation methods are used. Petri nets are one of these (new) basic methods. As discussed in § 1.1, the GNs can also be used as a means for simulating such (real) processes, but (in contrast with the Petri nets and by virtue of their transition condition predicates and characteristic functions) they can also include in themselves elements of analytical methods for description. For example, some characteristic functions can assign to a given token a value that is calculated by analytical functions.

"The NEFTOCHIM Petrochemical Combine (PCC) in Bourgas is one of the integrated industrial enterprises in Bulgaria which exerts a strong influence upon the dynamic development of the State Industrial Chemical Corporation, the economy of the country, and the international division of labour." [1]

The processes going in PCC can be modelled by GNs and the constructed models can be used for simulation of the processes in the combine. The models are investigated following the ideas for hierarchical models and models based on the union and the composition of other simpler GN-models introduced in § 1.1.

The modelling starts with the construction of a simplified global model of the combine in general. It has the from of Fig. 3.12.

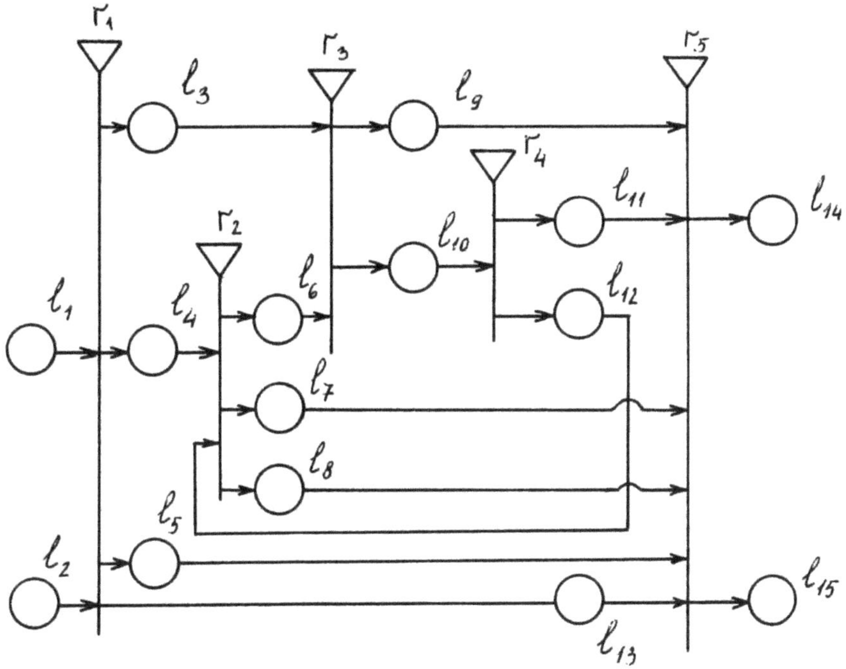

Fig. 3.12

The GN used can be, e.g., a second type intuitionistic fuzzy GN, i.e. a GN in which the tokens are replaced by "quantities" (see § 3 in App. 1) which "flow in the net. So a token ("quantity") enters place l_1 and this action represents the quantity of raw materials (oil) received. This place symbolizes Petrochemical Sea Port (Terminal) where distribution of raw material through the basic types of units is made: "chemical" (l_3) and refinery. The second one has the following two components: a "Catalitic Reforming" (l_4) and an "output of fuels" (l_5). A part of the production from l_4 is directed to l_6 and together with the production from l_3, is going (according to preliminary by set conditions) to l_9 (for production of plastics, fibres and rubber), or to l_{10} (for production of petrochemicals - in "petrochemical units"). A part of the "quality" (token) is going from the last place to place l_{11} (for production of solvents), and another part returns through place l_{12} to the "catalytic cracking unit" and there takes part with newly entered "qualities", in the production of fuels (l_7) and other products (l_8). The tokens (qualities) from places l_9, l_{11}, l_7, l_8, l_5 go to place l_{14} and that symbolizes the outcome of the finished product from the PCC.

By this part of the GN the movement (in general outline) of fuels, petrochemicals and plastics, produced and processed in PCC can be observed. Besides, the GN has a second (not differentiated separately) part that symbolizes the entry of information on the demand of petrochemicals (l_2), workout of this information (orders) (l_{13}) and report on the orders (l_{15}) - the effect that they are satisfied or not.

The tokens ("quantities") that go into l_1 have as initial characteristics the type of raw material, and during the time of movement in the GN they receive the following characteristics: "the kind and quantities of processing intermediate fractions" and "the kind and quantities of the finished product" depending on the type (kind) of places they enter. In the end in place l_{14} these tokens ("quantities") obtain the finished characteristic "working expenses" and other characteristics specified by the model's consumers.

The other type of tokens are going to place l_2 with the initial characteristic "information about the demand of goods with the necessary description, quantities, parameters, etc. and are leaving the GN with the final characteristic "the order is satisfied" or "the order is not satisfied" (as well as other data).

More detailed processing models can report on such factors as the moment when the raw material enters the PCC, the duration of processes, the moments when the products leave the PCC, the duration of waiting for processing of raw materials or fractions etc.

The other tokens can have their own priorities which determine the ways of the transfer. The transition condition predicates can have explicite forms. During simulation of the PPC-processes, different situations can occur in relation to the initial characteristics of the tokens in places l_1 and l_2, and also in relation with the characteristics of the tokens which are in the net at the initial time-moment (they correspond to the available quantities in the PPC).

The described GN reflects only the most global connections existing between the different units of PPC. When more details are required in the positions of the places and the transitions of the above net, new GNs will be standing. They will describe the separate subprocesses which are included in the most global processes.

Another situation is the one when a new GN (a subnet of the most global one) corresponds to a part of the places and transitions of the above GN, but does not cover their content.

A GN shown in Fig. 3.13 is an example, illustrating this. The components of this net correspond to places l_1, l_4, l_5 and the transition Z_1.

Briefly, the sense of the places of the new net are as follows:

l_1 - Petrochemical Sea Port (Terminal),

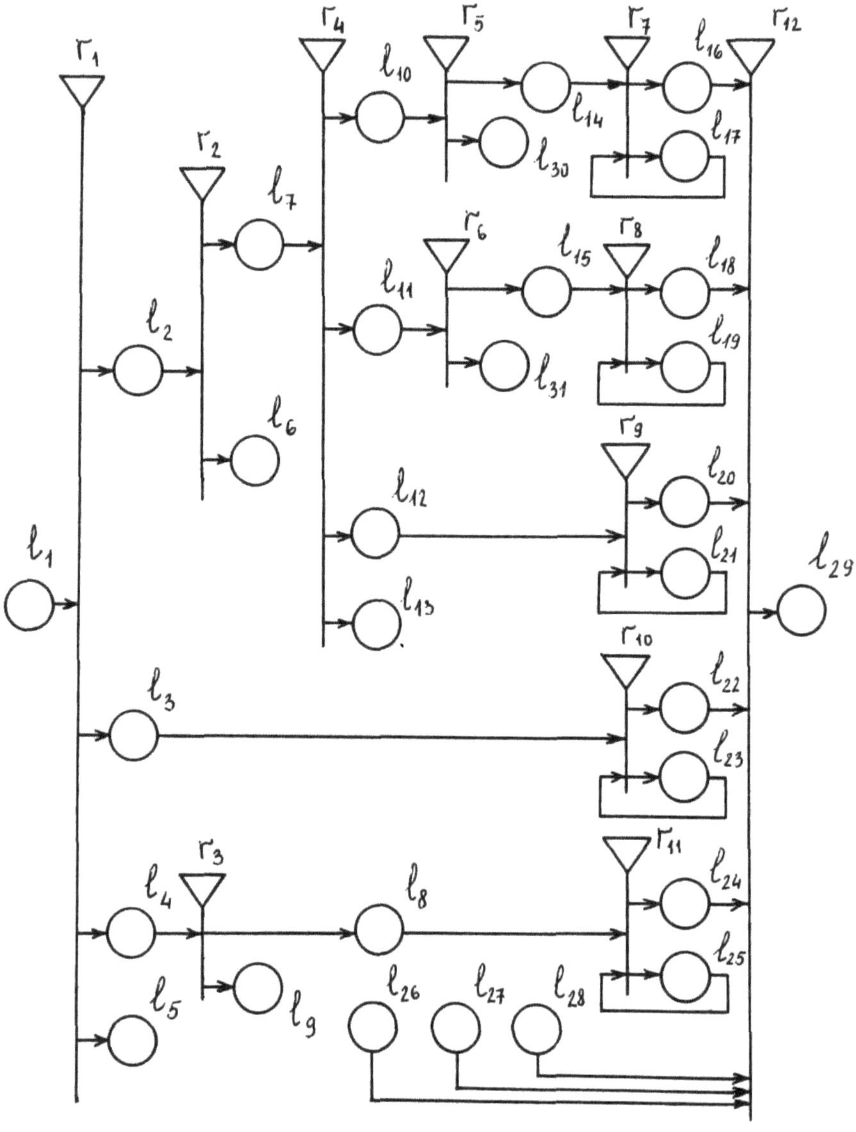

Γ_1
Γ_2
Γ_3
Γ_4
Γ_5
Γ_6
Γ_7
Γ_8
Γ_9
Γ_{10}
Γ_{11}
Γ_{12}
l_1
l_2
l_3
l_4
l_5
l_6
l_7
l_8
l_9
l_{10}
l_{11}
l_{12}
l_{13}
l_{14}
l_{15}
l_{16}
l_{17}
l_{18}
l_{19}
l_{20}
l_{21}
l_{22}
l_{23}
l_{24}
l_{25}
l_{26}
l_{27}
l_{28}
l_{29}
l_{30}
l_{31}

Fig. 3.13.

l_2 - production of fuel oil

$l_3, l_4, l_5, l_6, l_9, l_{13}, l_{30}$ - production of other cuts

l_7 - production of oil cut

l_{10}, l_{11} - production of cut C_4

l_{12} - production of high-octane gasoline

$l_{16}, l_{17}, \ldots, l_{25}$ - tanks

l_{26} - production of cut C_5

l_{27} - production of toluen.

The GN-models described here reflect the first steps of global modelling of the NEFTOCHIM Petrochemical Combine in Bourgas. In the proposed text, other aspects and methods for describing the petrochemical processes are not included.

REFERENCE:

[1] NEFTOCHIM Economic Combine-Bourgas, BulgarReklama Agency, Sofia, 1988.

§ 3.8: A SECOND TYPE OF INTUITIONISTIC FUZZY GENERALIZED NET-MODEL IN THE CHEMICAL INDUSTRY

Rumen Christov and Stoian Garbov

On the basis of the second type of intuitionistic fuzzy generalized nets (IFGN2) (see § 3 in App. 1) we shall describe a part of the technological scheme of the functioning of a silo-farm with pneumotransport in a plasticware company.

The raw materials are ferried with auto-cisterns and tanks. They are provided from 4 places and transferred to 10 store-bunkers (silos) or (if necessary) directly to the production lines. Raw materials must flow continuously to these lines. The resin enters 6 production lines from the silos by a distributor. Each line has an individual capacity and the material available for a definite time is consumed.

The IFGN2 describing this process has the form of Fig. 3.14. It has 3 transitions and 23 places with:

- places $l_1, \ldots, l_4$ - symbolising the accepting places;
- places $l_5, \ldots, l_{14}$ - symbolising the silos;
- place l_{15} - showing the transportation of the material to the production lines;
- place l_{17} - modeling the distributor;
- places $l_{18}, \ldots, l_{23}$ - symbolising the flow production lines.

All transitions have equal priority. The places l_{15} and l_{17} have the highest priority; the other places have equal priority. The transition conditions have the form:

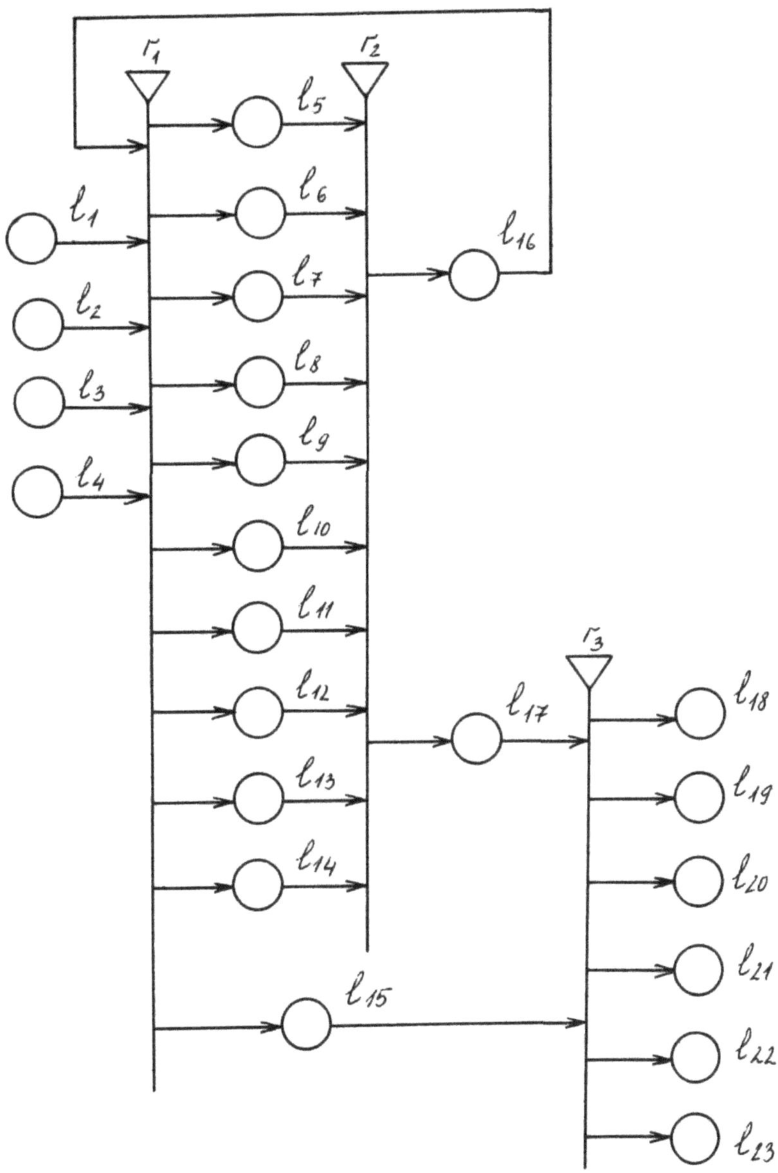
r_1
r_2
r_3
l_1
l_2
l_3
l_4
l_5
l_6
l_7
l_8
l_9
l_{10}
l_{11}
l_{12}
l_{13}
l_{14}
l_{15}
l_{16}
l_{17}
l_{18}
l_{19}
l_{20}
l_{21}
l_{22}
l_{23}

Fig. 3.14

r_1 =

	l_5	l_6	l_7	l_8	l_9	l_{10}	l_{11}	l_{12}	l_{13}	l_{14}	l_{15}
l_1	W_1	W_2	W_3	W_4	W_5	W_6	W_7	W_8	W_9	W_{10}	W_{11}
l_2	W_1	W_2	W_3	W_4	W_5	W_6	W_7	W_8	W_9	W_{10}	W_{11}
l_3	W_1	W_2	W_3	W_4	W_5	W_6	W_7	W_8	W_9	W_{10}	W_{11}
l_4	W_1	W_2	W_3	W_4	W_5	W_6	W_7	W_8	W_9	W_{10}	W_{11}
l_{16}	W_1	W_2	W_3	W_4	W_5	W_6	W_7	W_8	W_9	W_{10}	W_{11}

where

W_i = "the resin must go to the i-th silo" ($1 \le i \le 10$),

W_{11} = "the resin must go to the flow production lines, because there exists a flow production line with resin less than the critical quantity";

r_2 =

	l_{16}	l_{17}
l_5	W_{12}	W_{13}
l_6	W_{12}	W_{13}
l_7	W_{12}	W_{13}
l_8	W_{12}	W_{13}
l_9	W_{12}	W_{13}
l_{10}	W_{12}	W_{13}
l_{11}	W_{12}	W_{13}
l_{12}	W_{12}	W_{13}
l_{13}	W_{12}	W_{13}
l_{14}	W_{12}	W_{13}

where

W_{12} = "the resin must go to another silo";

W_{13} = "the resin must go to a production line with resin less than the critical quantity or with a minimal resin quantity";

r_3 =

	l_{18}	l_{19}	l_{20}	l_{21}	l_{22}	l_{23}
l_{15}	W_{14}	W_{15}	W_{16}	W_{17}	W_{18}	W_{19}
l_{17}	W_{14}	W_{15}	W_{16}	W_{17}	W_{18}	W_{19}

where

W_i = "the resin must go to the i-th production line which has the lowest resin quantity" and "the resin is below the critical limit" ($14 \le i \le 19$).

The characteristic functions Φ_i give the infinum and the supremum of the available resin quantity in the place l_i ($1 \le i \le 23$).

The IFGN2-models functioning starts at a certain time-moment T according to a certain absolute time-scale and the elementary time-step (t^o) being 2 minutes.

The features of this net are the discrete values of the predicates of the transition conditions but the net is an IFGN2, because the tokens are "quantities" which flow in the net and part of these quantities is lost in the process of transportation, i.e. there exists a degree of indetermination.

The following problems are solved (addaptively) by the model:

a) The optimal schedule of the raw materials ferrying is determined.

b) The optimal loading of the machines is determined.

c) The optimal capacity of the connected elements of the manufacturing is checked.

d) The possible waiting of the machines are determined.

On the basis of the IFGN2-model, the processes in a large class of productions of the chemical industry can be simulated and controlled (in real-time).

Chapter 4:

APPLICATIONS OF GENERALIZED NETS IN MEDICINE

The first research in the area of generalized net modelling of medical diagnosis dates back to 1983. In § 4.1 almost all the constructed nets are collected. The authors plan to realize all the described models by means of program package for generalized nets. A more general model of diagnostics is introduced in § 4.2.

§ 4.1: APPLICATIONS OF GENERALIZED NETS IN NEPHROLOGY

Joseph G. Sorsich and Krassimir T. Atanassov

In this chapter, an example of the application of GNs in medicine is given. Nephrology (see e.g. [1-16]), as a relatively new field in medicine, has been used to make things clearer and to demonstrate the possibilities of GNs. Some other applications in medicine are also available (e.g. [17]).

The mathematical foundations of our research are essentially different from other basic mathematical methods applicable in medicine (see e.g. [18-21]).

A renal damage is usually suspected in the course of a routine examination of a patient with some clinical signs and symptoms (dysuria, renal colic, edema, turbid urine), arterial hypertension or if some disturbances in blood sample or urinalyse are present (hyperazothemia, proteinuria, etc.). Based upon these common signs relating to kidney deseases, the following GN is constructed. This is not a definite model and it may be a subject of further modifications. New GNs, which will be the subnets of the general GN described below, may be added, and/or the subnets themselves may be modified. Our aim is also to demonstrate that such a subject in biological sciences as medicine is susceptible to comparatively simple and easy to understand mathematical formulation by GNs. The same may be realized in almost all other areas of medicine.

Initially, we shall describe one (general) GN which contains as subnets other GNs describing the process of diagnosing different signs and symptoms in nephrology. Let this GN be E. It has one input place L_1. The tokens of this GN will represent the patients of a nephrologic unit.

Everywhere we describe the transfer of one token (patient) or the tokens generated by it, but this is valid for every set of tokens (patients) transferring in the GN. At the end of this chapter we shall discuss these problems more widely, as well as the possibilities for application of this and similar GNs in different medical units.

The token goes to place L_1 with the initial characteristics "patient with probable kidney disease", next it splits into four tokens which represent the appropriate investigations of the patient. Later these tokens may be united into one token whose characteristics correspond to all the applied investigations. Part of these tokens may leave the GN if some of the signs are missing. The sense of the different GN's places are as follows:

L_2 - abnormal findings (changes) in urine are present,

L_3 - elevated blood urea nitrogen (BUN) is present,

L_4 - clinical signs and symptoms of renal disease are present,

L_5 - arterial hypertension is present,

L_6 - permanent proteinuria is present,

L_7 - hematuria is present,

L_8 - turbid urine is present,

L_9 - renal colic is present,

L_{10} - dysuria is present,

L_{11} - edematous patient is present. (Edemas of renal origin are always related to proteinuria, refer to § 9.1),

L_{12} - evaluation of arterial hypertension,

L_{13} - a specifying the type of renal damage,

L_{14} - arterial hypertension of renal (renoparenchimal) origin is suspected,

L_{15} - renovascular hypertension is suspected.

The transition conditions are as follows:

$$r_1 = \begin{array}{c|ccc} & L_2 & L_3 & L_4 \\ \hline L_1 & true & true & true \end{array};$$

$$r_2 = \begin{array}{c|ccc} & L_6 & L_7 & L_8 \\ \hline L_2 & W_1 & W_2 & W_3 \end{array}$$

where

W_1 = "permanent proteinuria is present",

W_2 = "hematuria is present",

W_3 = "turbid urine is present";

$$r_3 = \begin{array}{c|ccc} & L_9 & L_{10} & L_{11} \\ \hline L_4 & W_4 & W_5 & W_6 \end{array}$$

where

W_4 = "renal colic is present",

W_5 = "dysuria is present",

W_6 = "edematous patient is present";

$$r_4 = \begin{array}{c|cc} & L_{12} & L_{13} \\ \hline L_5 & true & true \end{array}$$

$$r_5 = \begin{array}{c|cc} & L_{12} & L_{13} \\ \hline L_{13} & W_7 & W_8 \end{array},$$

where

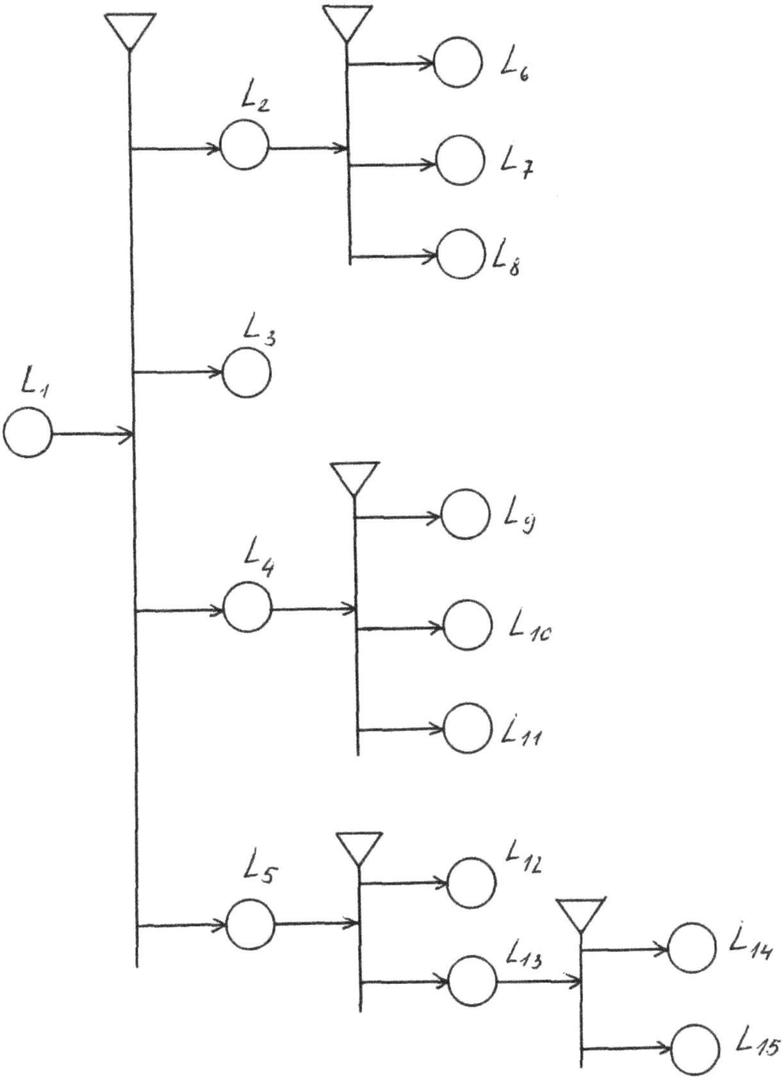
L1
L2
L3
L4
L5
L6
L7
L8
L9
L10
L11
L12
L13
L14
L15

W_7 = "there is suspicion of renoparenchimal arterial hypertension",

W_8 = "there is suspicion of renovascular arterial hypertension".

From the fact that a token (a unique one or one of many representations of a given token (patient) that enters GN E) enters places L_3, L_6, L_7, ..., L_{12}, L_{14}, L_{15}, it follows that this token is directed to the following subnets:

L_6 - to the GN from § 4.1.1,

L_7 - to the GN from § 4.1.2,

L_8 - to the GN from § 4.1.3,

L_3 - to the GN from § 4.1.4,

L_9 - to the GN from § 4.1.5,

L_{10} - to the GN from § 4.1.6,

L_{11} - to the GN from § 4.1.1,

L_{12} - to the GN from § 4.1.7,

L_{14} - to the GN from § 4.1.8,

L_{15} - to the GN from § 4.1.9.

On the basis of this model, we can achieve the following:

- control
- optimization
- simulation

of real processes.

Control is realized as follows: the physician describes the initial state of the patient (this is the initial characteristic of the token in L_1).

In the other offices, the token (patient) gets as current characteristic the results of the corresponding investigation. Thus on the basis of the total (at the moment) information about the patient (i. e. on the basis of all existing characteristics of the tokens, generated by the initial token (patient)), a decision for the subsequent direction of the tokens is determined (i. e. the subsequent patient's investigations).

When the transfer of more tokens (patients) in the net occurs, we can optimize their transfer with the aim of achieving minimum waiting-time in queues.

Knowing the probable distributions connected with the tranfer of the patients in different offices, on the basis of the total GN-model, the process can be simulated. Thus, different parameters of hospital units can be determined (e. g. workload of offices, average time for servicing a single patient and of a single office, average time-stay of a patient in a hospital unit etc.). This information can be used for optimal organization of different medical units.

Note, that direct use of our model is possible (in this form) in specialized (nephrological) units because only the processes of the nephrological disease diagnosis are described in it. If we want to apply similar models for other medical purposes (cardiology, neurology, obstetrics, laboratory investigations and their interpretation, etc.), we must include in it GN-models of the corresponding case. Some Bulgarian physicians are working on this problem.

There already exist more expert and/or informative medical systems to help physicians in diagnosis and therapeutics. Such expert and informative systems may be included in the framework of the described total GN-model or in some of its parts (subnets). For example, expert and/or informative systems for diagnostics can be used (e. g., in the frames of the global GN's memory (see App. 1) for calculation of the truth-values of some transition condition predicates and for calculation of some token's characteristics.

On the other hand, expert and informative systems for therapeutics may be included in the described GN-model, when the final token's characteristics are to be determined.

Below we describe other GN-models of basic nephrological signs and symptoms but these models have an independent sense because their entire inclusion in the GN E deprives them of the clearness which is easyly attainable by the GN-tools.

This paper is based on [22-30].

* * *

§ 4.1.1 Permanent proteinuria

Any permanent proteinuria in adults is a pathological situation. Proteinuria may be detected in different conditions. Usually it is asymptomatic, but the practitioner's first step must always be a careful physical examination of the patient - not only the urinary tract, but the cardiovascular system (including arterial blood pressure, eyeground) and other systems as well.

The tokens (patients) get into place l_1 with the initial characteristic "permanent proteinuria". Next they split and get in the sections with initial places l_2, l_3 and l_4 so that each gets the appropriate characteristic: "the result of the complete urinalysis", "plasma creatinine's level (or gromerular filtration rate)" and "X-ray examinations of the urinary tract". In the last case, the token leaves this GN and gets in the GN from § 4.1.2. The transition conditions r_1 and r_2 have the forms:

$$r_1 = \begin{array}{c|ccc} & l_2 & l_3 & l_4 \\ \hline l_1 & true & true & true \end{array}$$

$$r_2 = \begin{array}{c|cccc} & l_5 & l_6 & l_7 & l_8 \\ \hline l_2 & true & true & true & true \end{array}.$$

The token in place l_2 splits into four tokens which get into places l_5, l_6, l_7 and l_8 according to their characteristics: amount of 24-hours proteinuria, presence of glucosuria, microscopic investigation of urine sediment, urocultures.

The transition condition

$$r_3 = \begin{array}{c|cc} & l_9 & l_{10} \\ \hline l_3 & W_1 & W_2 \end{array}$$

where

W_1 = "finds out renal insufficiency",

$W_2 = \neg W_1$.

The token leaves this GN also and gets into the GN from § 4. 1. 4.

$$r_4 = \begin{array}{c|cc} & l_{11} & l_{12} \\ \hline l_5 & W_3 & W_4 \end{array}$$

where

W_3 = "proteinuria more that 3 gr, per 24 hours",

$W_4 = \neg W_3$.

$$r_5 = \begin{array}{c|cc} & l_{13} & l_{14} \\ \hline l_6 & W_5 & W_6 \end{array}$$

where

W_5 = "there is glucosuria"

(the token leaves this GN - the patient is directed to diabetologist),

$W_6 = \neg W_5$.

$$r_6 = \begin{array}{c|ccc} & l_{15} & l_{16} & l_{17} \\ \hline l_7 & true & true & true \end{array} .$$

The token in place l_7 splits into three tokens which get into places l_{15}, l_{16} and l_{17} with the corresponding characteristics: the results of the microscopic examination of urine sediment for white blood cells, red blood cells, bacteria.

The transition conditions r_7, r_8, ..., r_{13} have the forms

$$r_7 = \begin{array}{c|cc} & l_{18} & l_{19} \\ \hline l_8 & W_7 & W_8 \end{array}$$

wher

W_7 = "urine cultures are positive"

(the token leaves this GN and the patient is directed toward the GN described in § 4.1.3),

$W_8 = \neg W_7$;

$$r_8 = \begin{array}{c|cc} & l_{20} & l_{21} \\ \hline l_{11} & W_9 & W_{10} \end{array}$$

where

W_9 = "there is a paraproteinuria"

(the token leaves the GN - the patient is directed to hematologist),

$W_{10} = \neg W_9$;

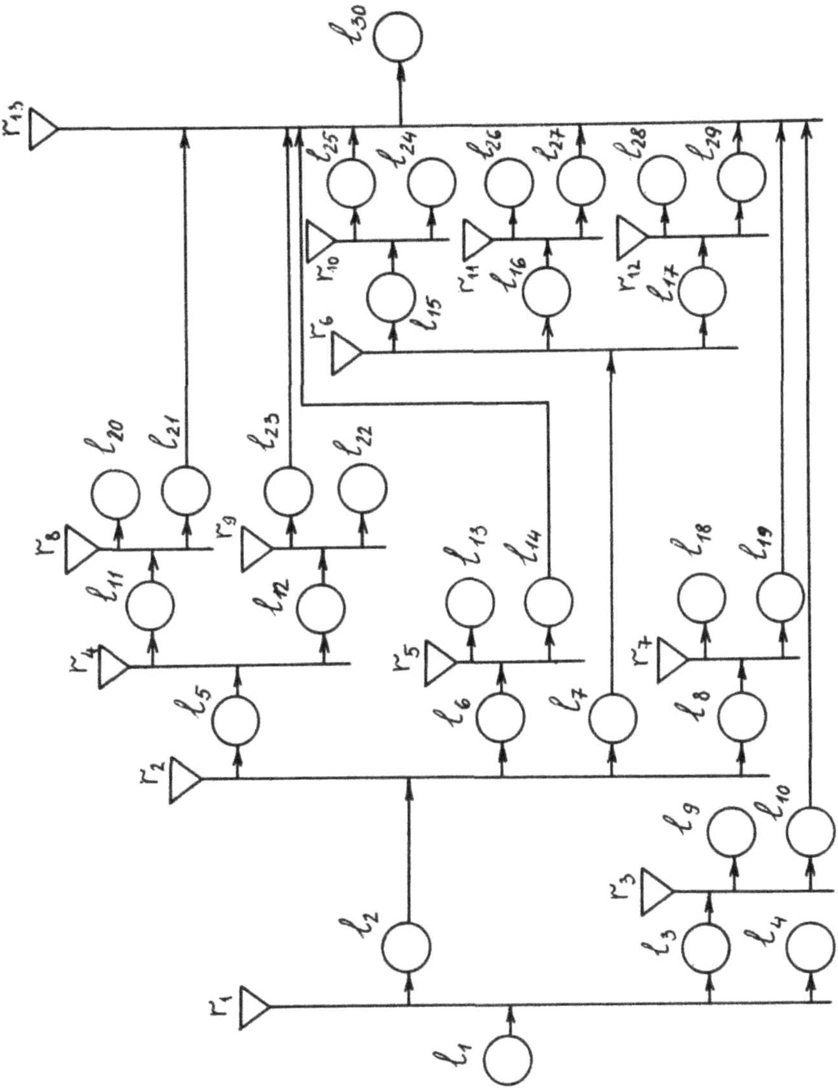
r1
r2
r3
r4
r5
r6
r7
r8
r9
r10
r11
r12
r13
l1
l2
l3
l4
l5
l6
l7
l8
l9
l10
l11
l12
l13
l14
l15
l16
l17
l18
l19
l20
l21
l22
l23
l24
l25
l26
l27
l28
l29
l30

$$r_9 = \begin{array}{c|cc} & l_{22} & l_{23} \\ \hline l_{12} & W_{11} & W_{12} \end{array}$$

where

W_{11} = "the proteinuria is combined with hematuria"

(the token goes to the GN from § 4.1.2);

$W_{12} = \neg W_{11}$;

$$r_{10} = \begin{array}{c|cc} & l_{24} & l_{25} \\ \hline l_{15} & W_{13} & W_{14} \end{array}$$

where

W_{13} = "leucocyturia is present";

$W_{14} = \neg W_{13}$;

$$r_{11} = \begin{array}{c|cc} & l_{26} & l_{27} \\ \hline l_{16} & W_{15} & W_{16} \end{array}$$

where

W_{15} = "hematuria is present";

$W_{16} = \neg W_{15}$;

$$r_{12} = \begin{array}{c|cc} & l_{28} & l_{29} \\ \hline l_{17} & W_{17} & W_{18} \end{array}$$

where

W_{17} = "bacteriuria is present";

$W_{18} = \neg W_{17}$;

$$r_{13} = \begin{array}{c|c} & l_{30} \\ \hline l_{21} & true \\ l_{23} & true \\ l_{14} & true \\ l_{25} & true \\ l_{27} & true \\ l_{29} & true \\ l_{19} & true \\ l_{10} & true \end{array}$$

If the token gets into place l_{24} or l_{28}, it enters the GN from § 4.1.3 and from place l_{26} to the GN from § 4.1.2.

Acute nephritic syndrome takes shape if in places l_5, l_9 and l_{26}, the tokens arising from an initial one are gathering (often the hematuria is macroscopic, the renal insufficiency is transitory, there are edema and arterial hypertension). The tokens in places l_{10}, l_{14}, l_{19}, l_{21}, l_{23}, l_{25}, l_{27} and l_{29} transit in place l_{30} - the patient is directed to a nephrologist, where complementary investigations are to be undertaken (immunological, renal biopsy, etc.).

* * *

§ 4.1.2 Hematuria

The presence of blood in urine (hematuria) is a common symptom in nephro-urology. The causal diagnosis sometimes is very complicated.

Hematuria may be isolated or combined with other signs and symptoms. Thus a complete examination of the patient is always indispensable.

The token gets into place l_1 with the initial characteristic - presence of hematuria, which is confirmed by a microscopic investigation of the urine sediment.

The transition condition

$$r_1 = \begin{array}{c|cc} & l_2 & l_2 \\ \hline l_1 & W_1 & W_2 \end{array}$$

where

W_1 = "there is a macroscopic (gross) hematuria";

$W_2 = \neg W_1$.

The token in place l_2 splits and gets into places l_4 and l_5 with the appropriate characteristics: results of the test with three glases and X-ray examinations of the urinary tract, because the transition condition is

$$r_2 = \begin{array}{c|ccc} & l_4 & l_5 & l_6 \\ \hline l_2 & true & true & true \end{array} .$$

Place l_6 is included in the framework of this GN as an illustration of one of the methods for constructing a general GN. It represents the relation between the GN from § 4.1.1 and the net presented here. In the framework of the general GN, tokens will transfer here from place l_4 of the GN from § 4.1.1 and this corresponds to transferring from one net to another.

$$r_4 = \begin{array}{c|ccc} & l_9 & l_{10} & l_{11} \\ \hline l_4 & W_3 & W_4 & W_5 \end{array}$$

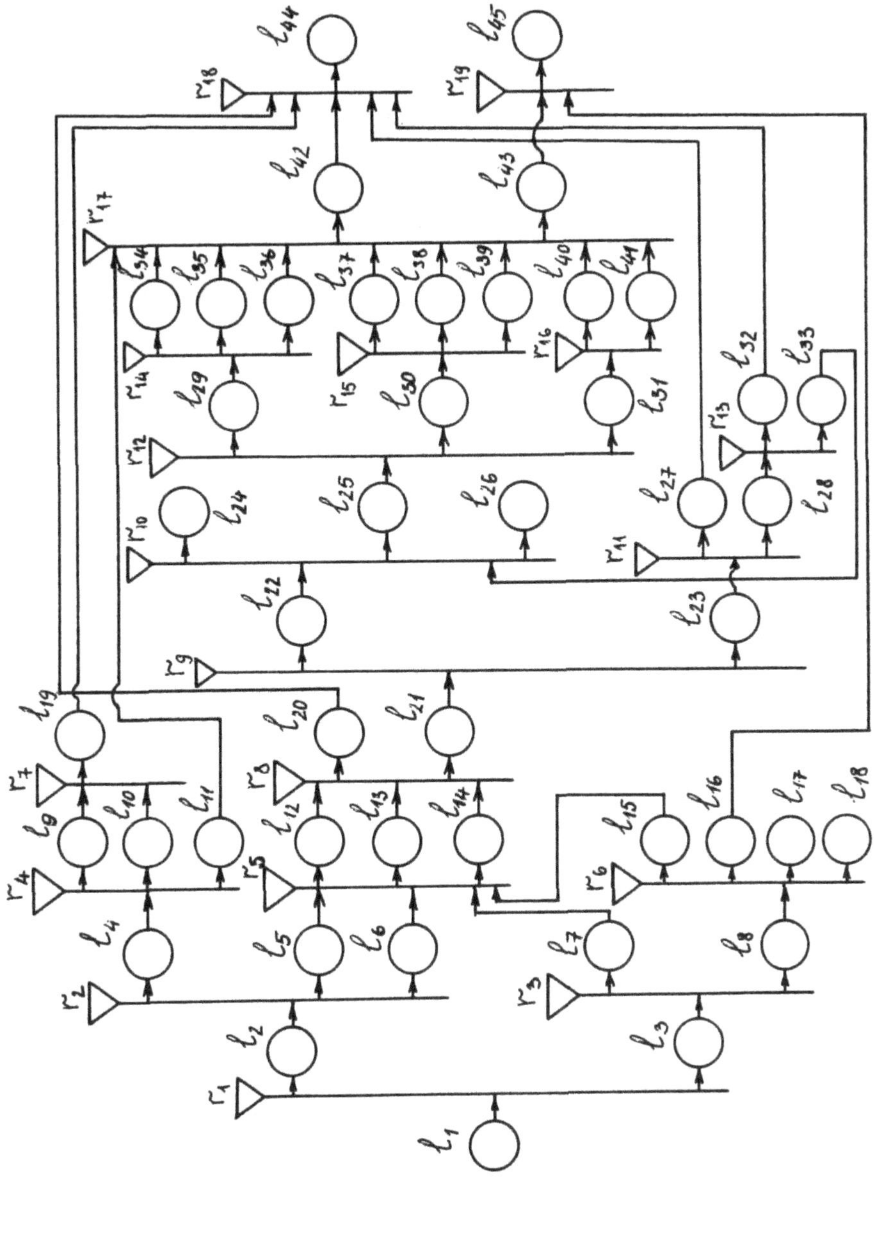
r1
r2
r3
r4
r5
r6
r7
r8
r9
r10
r11
r12
r13
r14
r15
r16
r17
r18
r19
l1
l2
l3
l4
l5
l6
l7
l8
l9
l10
l11
l12
l13
l14
l15
l16
l17
l18
l19
l20
l21
l22
l23
l24
l25
l26
l27
l28
l29
l30
l31
l32
l33
l34
l35
l36
l37
l38
l39
l40
l41
l42
l43
l44
l45

where

W_3 = "the presence of initial hematuria";

W_4 = "the presence of terminal hematuria";

W_5 = "the presence of total hematuria".

When the token gets into places l_9 and/or l_{10} the patient is directed to a urologist (l_{19}).

r_5 =

	l_{12}	l_{13}	l_{14}
l_5	true	W_6	W_7
l_6	true	W_6	W_7
l_7	true	W_6	W_7
l_{15}	true	W_6	W_7

where

W_6 = "there are no contraindications for intravenous urography";

$W_7 = \neg W_6$.

Thus this examination is performed.

Transition conditions r_7, r_8, r_9, r_{10}, r_{12}, r_{14}, r_{15} and r_{16} have the forms

r_7 =

	l_{19}
l_9	true
l_{10}	true

;

	l_{21}	l_{20}
$r_8 = l_{12}$	W_9	W_{10}
l_{13}	W_9	W_{10}
l_{14}	W_9	W_{10}

where

W_9 = "there are no urographically detectable changes or the results are uncertain";

$W_{10} = \neg W_9$

(in this case the patient is directed to a urologist);

	l_{22}	l_{23}
$r_9 = l_{20}$	W_{11}	W_{12}

where

W_{11} = "there are no cystoscopic changes or the diagnosis is uncertain";

$W_{12} = \neg W_{11}$;

	l_{24}	l_{25}	l_{26}
$r_{10} = l_{22}$	W_{13}	W_{14}	W_{15}
l_{33}	W_{13}	W_{14}	W_{15}

where

$W_{13} = \neg W_{14} \ \& \ \neg W_{15}$

(the token gets into place l_{24} - the patient is left under surveillance);

W_{14} = "there is a need for complementary X-ray examinations";

W_{15} = "renal biopsy is needed"

(the token in place l_{26} usually gets as characteristic a definite diagnosis);

$$r_{12} = l_{25} \mid \frac{l_{29} \quad l_{30} \quad l_{31}}{W_{16} \quad W_{17} \quad W_{18}}$$

where

W_{16} = "an extended examination of the renal parenchyma is needed";

W_{17} = "an examination of the bladder is needed";

W_{18} = "an extended examination of the vessels is needed" ;

$$r_{14} = l_{29} \mid \frac{l_{34} \quad l_{35} \quad l_{36}}{W_{19} \quad W_{20} \quad W_{21}}$$

where

W_{19} = "an arteriography is required";

W_{20} = "an sonography is required";

W_{21} = "a scanning is required";

$$r_{15} = l_{30} \mid \frac{l_{37} \quad l_{38} \quad l_{39}}{W_{22} \quad W_{23} \quad W_{24}}$$

where

W_{22} = "a retrograde cystoscopy is necessary";

W_{23} = "a retrograde ureteropyelography is necessary";

W_{24} = "an anterograde pyelography is necessary";

$$r_{16} = l_{31} \mid \frac{l_{40} \quad l_{41}}{W_{25} \quad W_{26}}$$

where

W_{25} = "an investigation of the renal arterial system is indispensable"

(aortography, selective renovasography),

W_{26} = "an investigation of the renal venous system is indispensable"

(cavography, renal phlebography).

The section of the GN described above ends by a transition with condition

$$r_{17} = \begin{array}{c|cc} & l_{42} & l_{43} \\ \hline l_{11} & W_{27} & W_{28} \\ l_{34} & W_{27} & W_{28} \\ l_{35} & W_{27} & W_{28} \\ l_{36} & W_{27} & W_{28} \\ l_{37} & W_{27} & W_{28} \\ l_{38} & W_{27} & W_{28} \\ l_{39} & W_{27} & W_{28} \\ l_{40} & W_{27} & W_{28} \\ l_{41} & W_{27} & W_{28} \end{array}$$

where

W_{27} = "urologic cause of hematuria is found"

(the patient is directed to a urologist - l_{44});

W_{28} = "a nephrologic cause of a hematuria is found".

(the patient is directed to nephrologist - l_{45}).

Transition conditions r_9 and r_{13} have the forms

$$r_9 = \begin{array}{c|cc} & l_{27} & l_{28} \\ \hline l_{23} & W_{29} & W_{30} \end{array}$$

where

W_{29} = "bleeding from the lower urinary tract is discovered"

(and the token gets into place l_{44});

W_{30} = "bleeding from the upper urinary tract is discovered";

$$r_{13} = \begin{array}{c|cc} & l_{32} & l_{33} \\ \hline l_{28} & W_{31} & W_{32} \end{array}$$

where

W_{31} = "the bleeding is unilateral",

(and the token gets into place l_{44}),

$W_{32} = \neg W_{31}$

(the token gets into place l_{33} and the condition r_{10} is verified).

Transition conditions r_3 and r_6 have the forms

$$r_3 = \begin{array}{c|cc} & l_{7} & l_{8} \\ \hline l_{3} & W_{33} & W_{34} \end{array}$$

where

W_{33} = "presence of hematuria combined with proteinuria that cannot be explained by the presence of blood in urine only";

W_{34} = "presence of isolated hematuria";

$$r_6 = \begin{array}{c|cccc} & l_{15} & l_{16} & l_{17} & l_{18} \\ \hline l_{8} & W_{35} & W_{36} & W_{37} & W_{38} \end{array}$$

where

W_{35} = "X-ray examination is needed"

(and a token's transfer to a 5-th transition follows);

W_{36} = "there are definite signs of nephropathy"

(the patient is directed to a nephrologist - l_{45});

W_{37} = "investigations of the concomitant proteinuria is needed"

(the token leaves this GN and gets in a GN from § 4.1.1),

W_{38} = "investigation of the concomitant arterial hypertension is needed"

(the token leaves this GN and gets into place L_5 of the basic GN E and from there it goes to some of the GNs from § 4.1.7, § 4.1.8 or § 4.1.9).

*
* *

§ 4.1.3 Turbid urine

Here some advantages of GNs and their applications in the initial steps of identification and investigation of a turbid urine sample will be illustrated.

Tokens (patients) enter place l_1 with the initial characteristic "complains of voiding turbid urine". The tokens' transfer from place l' to place l" shall be denoted by l' -> l".

The transition condition of the GN are

$$r_1 = \begin{array}{c|c} & l_2 \\ \hline l_1 & true \end{array} ;$$

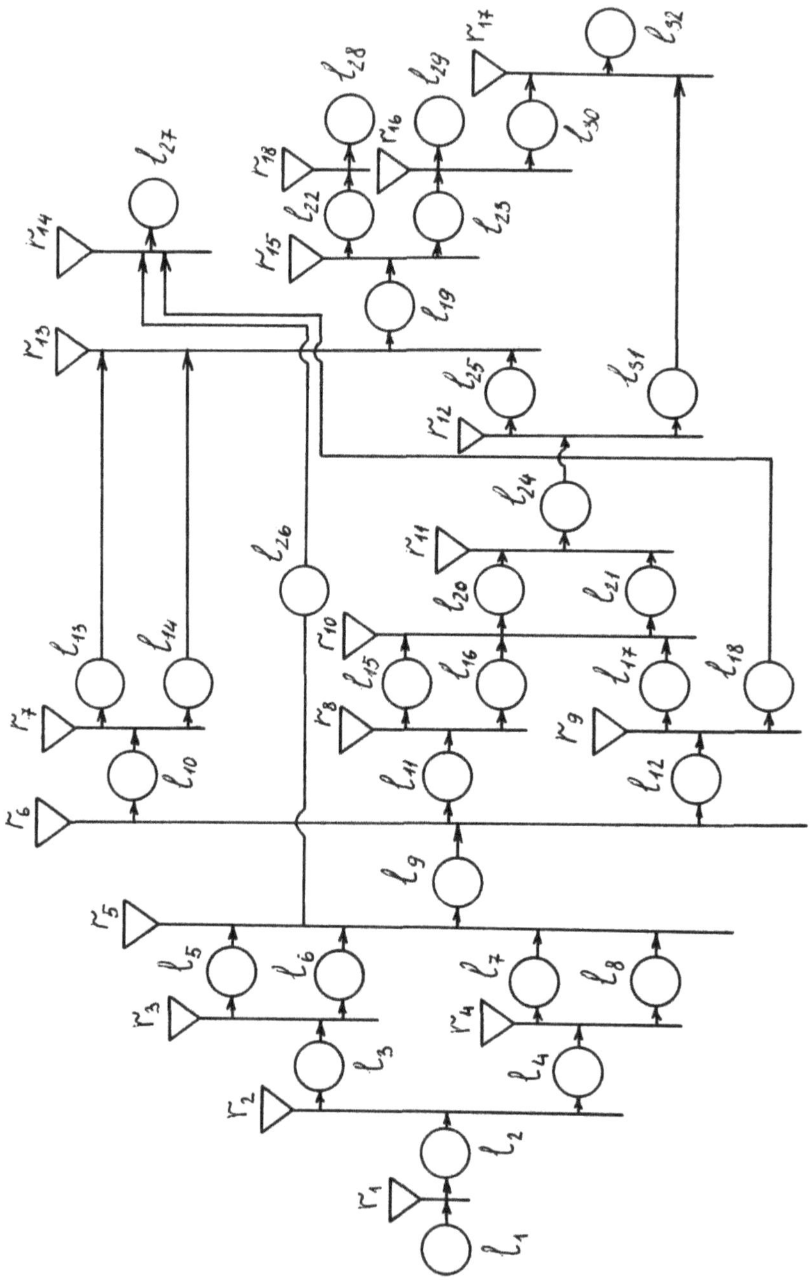

r1
r2
r3
r4
r5
r6
r7
r8
r9
r10
r11
r12
r13
r14
r15
r16
r17
r18
l1
l2
l3
l4
l5
l6
l7
l8
l9
l10
l11
l12
l13
l14
l15
l16
l17
l18
l19
l20
l21
l22
l23
l24
l25
l26
l27
l28
l29
l30
l31
l32

$$r_2 = \begin{array}{c|cc} & l_2 & l_3 \\ \hline l_2 & W_1 & W_2 \end{array}$$

where

W_1 = "acidification of the urine specimen is performed";

W_2 = "investigation of urine with test strip";

$$r_3 = \begin{array}{c|cc} & l_5 & l_6 \\ \hline l_3 & W_3 & W_4 \end{array}$$

where

W_3 = "the turbidity of urine disapears";

$W_4 = \neg W_3$;

$$r_4 = \begin{array}{c|cc} & l_7 & l_8 \\ \hline l_4 & W_5 & W_6 \end{array}$$

where

W_5 = "there are no changes in the test strip";

$W_6 = \neg W_5$;

$$r_5 = \begin{array}{c|cc} & l_9 & l_{26} \\ \hline l_5 & W_7 & W_8 \\ l_6 & true & false \\ l_7 & W_9 & W_{10} \\ l_8 & true & false \end{array}$$

where

W_7 = "the transfer $l_4 \rightarrow l_8$ is accomplished ";

$W_8 = \neg W_7$;

W_9 = "the transfer $l_3 \rightarrow l_6$ is accomplished ";

$W_{10} = \neg W_9$;

$$r_6 = l_9 \left| \frac{l_{10} \quad l_{11} \quad l_{12}}{W_{11} \quad W_{12} \quad W_{13}} \right.$$

where

W_{11} = "bacteriological investigation of urine is undertaken";

W_{12} = "microscopic urine sediment examination is carried out";

W_{13} = "macroscopic examination of urine is accomplished";

$$r_7 = l_{10} \left| \frac{l_{13} \quad l_{14}}{W_{14} \quad W_{15}} \right.$$

where

W_{14} = "there is bacteriuria";

$W_{15} = \neg W_{14}$;

$$r_8 = l_{11} \left| \frac{l_{15} \quad l_{16}}{W_{16} \quad W_{17}} \right.$$

where

W_{16} = "there is leucocyturia";

$W_{17} = \neg W_{16}$;

$$r_9 = l_{12} \left| \frac{l_{17} \quad l_{18}}{W_{18} \quad W_{19}} \right.$$

where

W_{18} = "there are urethral filaments in the urine specimen",

$W_{19} = \neg W_{18}$;

$$r_{10} = \begin{array}{c|cc} & l_9 & l_{26} \\ \hline l_{15} & W_{20} & false \\ l_{16} & false & W_{21} \\ l_{17} & W_{20} & W_{21} \end{array}$$

where

W_{20} = "the transfer $l_{11} \to l_{15}$ is accomplished";

$W_{21} = \neg W_{20}$;

$$r_{11} = \begin{array}{c|c} & l_9 \\ \hline l_{20} & true \\ l_{21} & true \end{array};$$

$$r_{12} = \begin{array}{c|cc} & l_{25} & l_{31} \\ \hline l_{24} & W_{22} & W_{23} \end{array}$$

where

W_{22} = "there is no prostatic hypertrophy";

$W_{23} = \neg W_{22}$;

$$r_{13} = \begin{array}{c|c} & l_{19} \\ \hline l_{13} & true \\ l_{14} & true \\ l_{25} & true \end{array};$$

$$r_{14} = \frac{l_{27}}{\begin{array}{l|l} l_{26} & \text{true} \\ l_{18} & \text{true} \end{array}};$$

$$r_{15} = \frac{l_{22} \quad l_{23}}{l_{19} \mid W_{24} \quad W_{25}}$$

where

W_{24} = "there are no X-ray alternations in the kidney and urinary tract";

$W_{25} = \neg W_{24}$;

$$r_{16} = \frac{l_{29} \quad l_{30}}{l_{23} \mid W_{26} \quad W_{27}}$$

where

W_{26} = "X-ray alternations in the urinary tract are demonstrated" & $\neg W_{27}$;

W_{27} = "X-ray alternations in the kidney are present".

The transfer $l_1 \rightarrow l_2$ means that it is indispensable to investigate a fresh urine specimen. The transfer $l_5 \rightarrow l_{26}$ means that the turbidity of urine disapears and thus the investigation is discontinued. The transfers $l_5 \rightarrow l_9$, $l_7 \rightarrow l_9$ and $l_8 \rightarrow l_9$ impose an extension of investigations. The transfers $l_{15} \rightarrow l_{20}$ and $l_{17} \rightarrow l_{20}$ are simultaneously performed, which means that leucocyturia and urethral filaments are found. The transfers $l_{16} \rightarrow l_{21}$ and $l_{17} \rightarrow l_{21}$, which are also simultaneously performed, signify the absent leucocyturia, but there are urethral filaments. The token in place l_{24} takes ascharacteristic the result of the investigation of the prostatic gland, and in place l_{19}, the characteristics of the results from the intravenouse urography. The transfers $l_{26} \to l_{27}$ and $l_{18} \to l_{27}$ mean the end of the investigation; $l_{23} \to l_{29}$ means that the patient is directed to a specialised nephrologic unit; $l_{30} \to l_{32}$ and $l_{31} \to l_{32}$ means that the patient is directed to urological unit; $l_{22} \to l_{28}$ means that the patient remains under observation after appropriate therapy.

* * *

§ 4.1.4 Elevated blood urea nitrogen

Estimation of the Glomerular Filtration Rate (GFR) is of great importance both in the initial evaluation and in subsequent the follow-up of patients with renal disease. Measurement of blood urea nitrogen (BUN) is probably the most commonly used, albeit the least accurate, clinical method for estimating GFR. The serum creatinine concentration provides a more reliable estimate of GFR.

In place l_1 we have a token (patient) with the initial characteristic "elevated BUN". Next it splits and gets in places l_2, l_3 and l_4, as the transition condition r_1 has the form

$$r_1 = \begin{array}{c|ccc} & l_2 & l_3 & l_4 \\ \hline l_1 & true & true & true \end{array}.$$

Each place gets an appropriate characteristics "results of hemodynamics investigation" (l_2); "state of hydratation - evaluation of

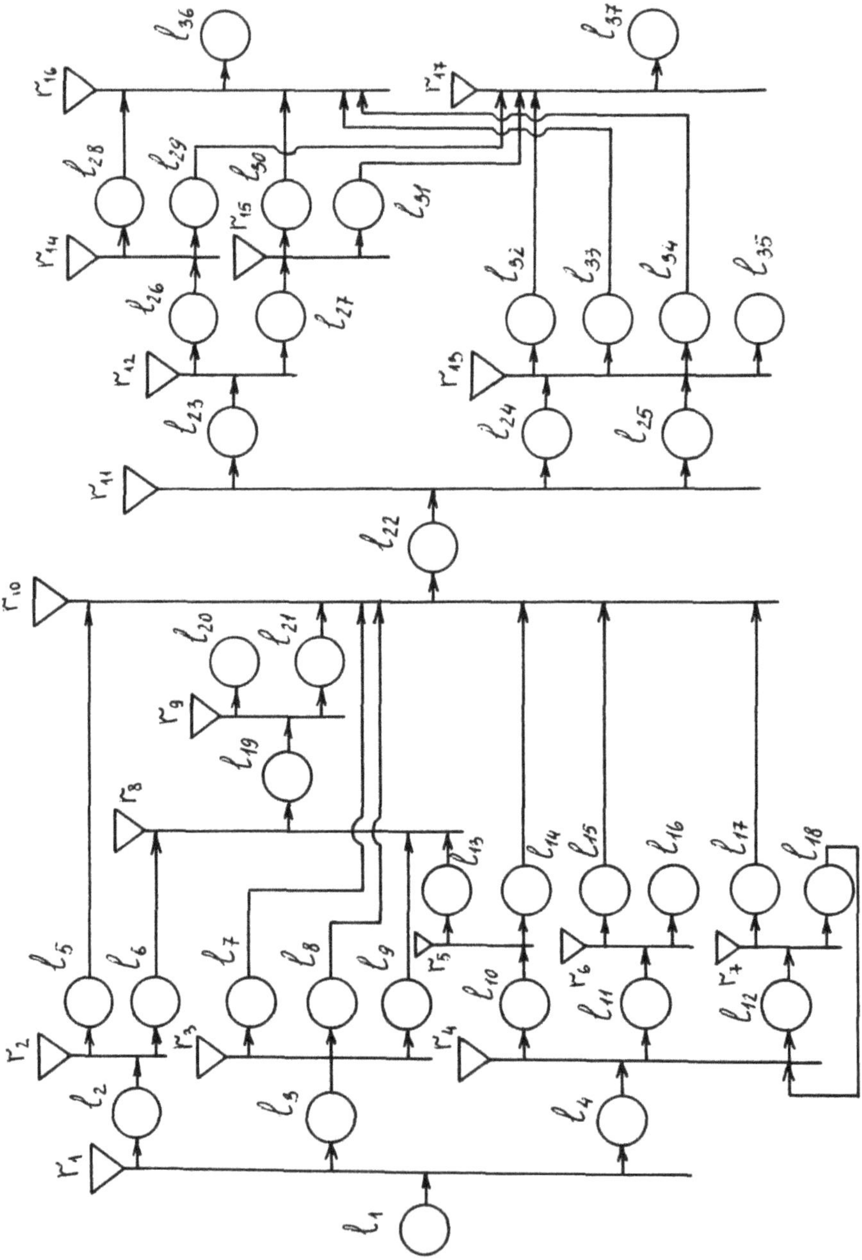

r_1 r_2 r_3 r_4 r_5 r_6 r_7 r_8 r_9 r_{10} r_{11} r_{12} r_{13} r_{14} r_{15} r_{16} r_{17}
ℓ_1 ℓ_2 ℓ_3 ℓ_4 ℓ_5 ℓ_6 ℓ_7 ℓ_8 ℓ_9 ℓ_{10} ℓ_{11} ℓ_{12} ℓ_{13} ℓ_{14} ℓ_{15} ℓ_{16} ℓ_{17} ℓ_{18} ℓ_{19} ℓ_{20} ℓ_{21} ℓ_{22} ℓ_{23} ℓ_{24} ℓ_{25} ℓ_{26} ℓ_{27} ℓ_{28} ℓ_{29} ℓ_{30} ℓ_{31} ℓ_{32} ℓ_{33} ℓ_{34} ℓ_{35} ℓ_{36} ℓ_{37}

extracellular volume" (l_3), and "urinalysis" (l_4).

The transition condition

$$r_2 = \begin{array}{c|cc} & l_5 & l_6 \\ \hline l_2 & W_1 & W_2 \end{array},$$

where

W_1 = "there is a reduction in effective arterial blood volume";

$W_2 = \neg W_1$.

The transition condition

$$r_3 = \begin{array}{c|ccc} & l_7 & l_8 & l_9 \\ \hline l_3 & W_3 & W_4 & W_5 \end{array},$$

where

W_3 = "there are symptoms of hyperhydratation";

W_4 = "normal findings";

W_5 = "dehydration is present".

The transition condition

$$r_4 = \begin{array}{c|ccc} & l_{10} & l_{11} & l_{12} \\ \hline l_4 & true & true & true \\ l_{18} & true & true & true \end{array}.$$

From place l_4 the token splits and gets into places l_{10} (determination of urea in the urine), l_{11} (evaluation for the presence of protein), l_{12} (examination of urine sediment for broad waxy casts).

The transition conditions r_5, r_6 and r_7 have the following forms:

$$r_5 = \begin{array}{c|cc} & l_{13} & l_{14} \\ \hline l_{10} & W_6 & W_7 \end{array},$$

where

W_6 = "there is an increase of urea excretion in urine";

W_7 = "there is a decrease of urine urea" ;

$$r_6 = \begin{array}{c|cc} & l_{15} & l_{16} \\ \hline l_{11} & W_8 & W_9 \end{array}$$

where

W_8 = "proteinuria is present" (see § 4.1.3);

$W_9 = \neg W_8$.

$$r_7 = \begin{array}{c|cc} & l_{17} & l_{18} \\ \hline l_{12} & W_{10} & W_{11} \end{array}$$

where

W_{10} = "broad casts are present";

$W_{11} = \neg W_{10}$

and this investigation may be repeated (see § 4.1.3).

The tokens from places l_5, l_9 and l_{13} get into place l_{19} (there are signs of functional, reversible renal failure).

The transition conditions r_8 and r_9 have the following forms:

$$r_8 = \begin{array}{c|c} & l_{19} \\ \hline l_6 & true \\ l_9 & true \\ l_{13} & true \end{array}$$

$$r_9 = \begin{array}{c|cc} & l_{20} & l_{21} \\ \hline l_{19} & W_{12} & W_{13} \end{array}$$

where

W_{12} = "BUN is normalized after an appropriate therapy";

$W_{13} = \neg W_{12}$.

The tokens from places l_6, l_7, l_8, l_{14}, l_{15}, l_{17} and l_{21} get into place l_{22} because organic renal failure is suspected.

The transition conditions r_{10} and r_{11} have the following forms:

$$r_{10} = \begin{array}{c|c} & l_{22} \\ \hline l_5 & true \\ l_{21} & true \\ l_7 & true \\ l_8 & true \\ l_{14} & true \\ l_{15} & true \\ l_{17} & true \end{array}$$

$$r_{11} = \begin{array}{c|ccc} & l_{23} & l_{24} & l_{25} \\ \hline l_{22} & true & true & true \end{array}.$$

Tokens from place l_{22} must be submitted to further investigations: blood (l_{23}), ultrasonographic (l_{24}), and radiologic (l_{25}) evaluations of kidneys.

The transition conditions r_{12}, r_{13}, r_{14} and r_{15} have the following forms:

$$r_{12} = \begin{array}{c|cc} & l_{26} & l_{27} \\ \hline l_{23} & true & true \end{array}$$

$$r_{13} = \begin{array}{c|cccc} & l_{32} & l_{33} & l_{34} & l_{35} \\ \hline l_{24} & W_{14} & W_{15} & W_{16} & W_{17} \\ & & & & \\ l_{25} & W_{14} & W_{15} & W_{16} & W_{17} \end{array}$$

where

W_{14} = "the size of both kidneys is normal";

W_{15} = "presence of bilateral small kidneys";

W_{16} = "the kidneys are enlarged";

W_{17} = "there is enlargement of the pyelocalicial system";

and urologist must be consulted because obstructive uropathy is suspected;

$$r_{14} = \begin{array}{c|cc} & l_{28} & l_{29} \\ \hline l_{26} & W_{18} & W_{19} \end{array}$$

where

W_{18} = "anaemia is present";

$W_{19} = \neg W_{18}$;

$$r_{15} = \begin{array}{c|cc} & l_{30} & l_{31} \\ \hline l_{27} & W_{20} & W_{21} \end{array}$$

where

W_{20} = "there is hypocalcemia";

$W_{21} = \neg W_{20}$;

$$r_{16} = \begin{array}{c|c} & l_{36} \\ \hline l_{28} & true \\ l_{30} & true \\ l_{33} & true \\ l_{34} & true \end{array}$$

$$r_{17} = \begin{array}{c|c} & l_{37} \\ \hline l_{29} & true \\ l_{31} & true \\ l_{32} & true \end{array} .$$

Tokens from places l_{28}, l_{30}, l_{33}, and l_{34} get into place l_{36} with the characteristics of chronic kidney failure. From places l_{29}, l_{31} and l_{32} the tokens get into place l_{37} with the characteristic of acute kidney failure.

The above constructed GN describes the diagnostic process when a patient presents high level of BUN.

*
* *

§ 4.1.5 Renal colic

The diagnosis of renal colic is modelled in the framework of this GN. The tokens (patients) get into place l_1 with the initial characteristic "symptoms of renal colic". Then they split up and get into the sections with initial places l_2 and l_3 and the appropriate characteristics - investigation of urine specimen and the result of the symptomatic treatment (the subside of renal colic), because of the transition condition

$$r_1 = \begin{array}{c|cc} & l_2 & l_3 \\ \hline l_1 & true & true \end{array} .$$

Depending on the values of these characteristics they continue their way in the GN. The transition condition

$$r_2 = \begin{array}{c|ccc} & l_4 & l_5 & l_6 \\ \hline l_2 & true & true & true \\ l_{31} & true & true & true \end{array} .$$

The tokens in places l_2 and l_{31} split and subsequently get into places l_4, l_5 and l_6, with the corresponding their characteristics: the results of macroscopic, test-strip and cytobacteriologic examinations. They advance further depending on the values of the transition conditions r_3, r_4, r_5 and r_6:

$$r_3 = \begin{array}{c|cccc} & l_7 & l_8 & l_9 & l_{10} \\ \hline l_4 & W_1 & W_2 & W_3 & W_4 \end{array}$$

where

W_1 = "the urine is without any microscopic change";

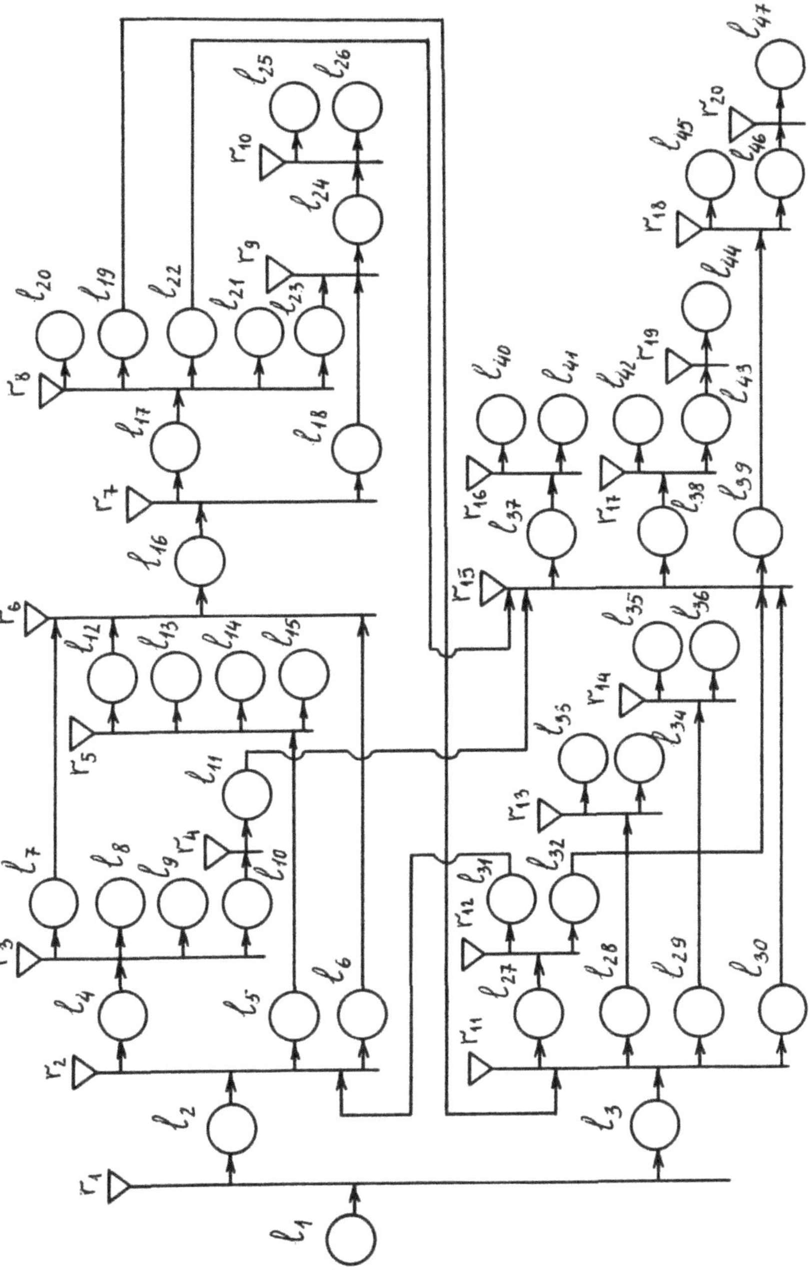
l1
r1
l2
l3
r2
l4
l5
l6
r3
l7
l8
l9
r4
l10
l11
r5
l12
l13
l14
l15
r6
l16
r7
l17
l18
r8
l19
l20
l21
l22
l23
r9
l24
r10
l25
l26
r11
l27
l28
l29
l30
r12
l31
l32
r13
l33
l34
r14
l35
l36
r15
l37
l38
l39
r16
l40
l41
r17
l42
l43
r19
l44
r18
l45
l46
r20
l47

W_2 = "there are hematuria and/or blood coagulums"

(the token enters the GN from § 4.1.2);

W_3 = "the urine sample is turbid"

(the token enters the GN from § 4.1.3);

W_4 = "a calculus is emitted";

$$r_4 = \begin{array}{c|c} & l_{11} \\ \hline l_{10} & true \end{array}$$

$$r_5 = \begin{array}{c|cccc} & l_{12} & l_{13} & l_{14} & l_{15} \\ \hline l_5 & W_5 & W_6 & W_7 & W_8 \end{array}$$

where

W_5 = "there are no changes";

W_6 = "there is proteinuria"

(the token enters the GN from § 4.1.1);

W_7 = "there is blood in urine"

(the token enters the GN from § 4.1.2);

W_8 = "nitrites are present";

$$r_6 = \begin{array}{c|c} & l_{16} \\ \hline l_7 & true \\ l_{12} & true \\ l_5 & true \end{array} .$$

The transfer of tokens into place l_{16} means that a cytobacteriological examination is performed, which is a parallel investigation of the urine's sediment (l_{17}) and microbiology (l_{18}) because

$$r_7 = \begin{array}{c|cc} & l_{17} & l_{18} \\ \hline l_{16} & true & true \end{array}.$$

The transition conditions r_8, r_9 and r_{10} have the forms

$$r_8 = \begin{array}{c|ccccc} & l_{19} & l_{20} & l_{21} & l_{22} & l_{23} \\ \hline l_{17} & W_9 & W_{10} & W_{11} & W_{12} & W_{13} \end{array},$$

where

W_9 = "the urine sediment is normal";

W_{10} = "there is hematuria"

(the token enters the GN from § 4.1.2);

W_{11} = "there is leucocyturia"

(the token enters the GN from § 4.1.3);

W_{12} = "there is crystaluria";

W_{13} = "there is bacteriuria";

$$r_9 = \begin{array}{c|c} & l_{24} \\ \hline l_{23} & true \end{array};$$

$$r_{10} = \begin{array}{c|cc} & l_{25} & l_{26} \\ \hline l_{24} & W_{14} & W_{15} \end{array}$$

where

W_{14} = "there is no bacteriuria";

$W_{15} = \neg W_{14}$.

The token's characteristic in place l_{25} is "no urinary tract infection is present" and in place l_{26}, "appropriate therapy must be undertaken".

The tokens in places l_3 and l_{22} split and get on parallel ways.

The transition conditions r_{11}, r_{12}, r_{13} and r_{14} have the forms

$$r_{11} = \begin{array}{c|cccc} & l_{27} & l_{28} & l_{29} & l_{30} \\ \hline l_3 & true & true & true & true \\ l_{19} & true & true & true & true \end{array}$$

$$r_{12} = \begin{array}{c|cc} & l_{31} & l_{32} \\ \hline l_{27} & W_{16} & W_{17} \end{array}$$

where

W_{16} = "the transfer $l_3 \to l_{27}$ is previously accomplished";

$W_{17} = \neg W_{16}$;

$$r_{13} = \begin{array}{c|cc} & l_{33} & l_{34} \\ \hline l_{28} & W_{18} & W_{19} \end{array}$$

where

W_{18} = "there is no decrease in diuresis";

$W_{19} = \neg W_{18}$

(the token enters the GN from § 4.1.4);

$$r_{14} = \begin{array}{c|cc} & l_{35} & l_{36} \\ \hline l_{29} & W_{20} & W_{21} \end{array}$$

where

W_{18} = "there is no abnormal increase in body temperature";

$W_{21} = \neg W_{20}$.

The transition condition

$$r_{15} = \begin{array}{c|ccc} & l_{37} & l_{38} & l_{39} \\ \hline l_{22} & true & true & true \\ l_{11} & true & true & true \\ l_{32} & true & true & true \\ l_{30} & true & true & true \end{array}.$$

Therefore, when a token gets into place l_{11}, l_{22}, l_{30} or l_{32} it splits into three. The transition conditions r_{16}, r_{17}, ..., r_{20} have the forms:

$$r_{16} = \begin{array}{c|cc} & l_{40} & l_{41} \\ \hline l_{37} & W_{22} & W_{23} \end{array}$$

where

W_{22} = "the renal function is normal";

$W_{23} = \neg W_{22}$

(the token enters the GN from § 4.1.4);

$$r_{17} = \begin{array}{c|cc} & l_{42} & l_{43} \\ \hline l_{38} & W_{24} & W_{25} \end{array}$$

where

W_{24} = "the X-ray of the urinary tract is normal";

$W_{25} = \neg W_{24}$;

$$r_{18} = \begin{array}{c|cc} & l_{45} & l_{46} \\ \hline l_{39} & W_{26} & W_{27} \end{array}$$

where

W_{26} = "the metabolic functional tests are normal";

$W_{27} = \neg W_{26}$;

$$r_{19} = \begin{array}{c|c} & l_{44} \\ \hline l_{43} & true \end{array}$$

$$r_{20} = \begin{array}{c|c} & l_{47} \\ \hline l_{46} & true \end{array} .$$

When tokens get into places l_8, l_9, l_{13}, l_{14}, l_{20}, l_{21}, l_{34}, l_{36} and l_{41}, they leave this GN and get to the GN from § 4.1.4.

In places l_{25}, l_{33}, l_{35}, l_{42} and l_{45} they end their movement in the net with the characteristic "normal findings", and tokens in places l_{26}, l_{44} and l_{47} end with the characteristic "directing towards appropriate treatment".

*
* *

§ 4.1.6 Dysuria

In normal micturition cycle, the continence phase depends on the ability of the bladder adapt itself to urine volume and maintain pressure in urethra. During micturition, the bladder contraction associated with sphincter relaxation empties the bladder. Dysuria denotes difficulty or pain associated with voiding. It may result from a wide variety of pathological conditions and it is generally due to a cervico-urethro-prostatic obstacle. When this it is not the case one should investigate for a functional obstruction of the striated sphincter and the cervix, or a bladder hypocontractility induced by denervation, inhibition or muscular insufficiency.

The tokens (patients) get into place l_1 with the initial characteristic "dysuria". Next they split up and get into places l_2, l_3 and l_4 so that in each place they get the characteristics: "results from the case history" (l_2), "general examination" (l_3) or "complementary laboratory investigations" (l_4), because

$$r_1 = \begin{array}{c|ccc} & l_2 & l_3 & l_4 \\ \hline l_1 & true & true & true \end{array}.$$

The transition conditions r_2, r_5 and r_4 have the forms

$$r_2 = \begin{array}{c|cc} & l_5 & l_6 \\ \hline l_2 & W_1 & W_2 \\ l_3 & W_1 & W_2 \\ l_{17} & false & true \\ l_{55} & false & true \end{array}$$

where

W_1 = "there is a certain disease that may cause dysuria";

$W_2 = \neg W_1$;

$$r_5 = \begin{array}{c|ccc} & l_9 & l_{10} & l_{11} \\ \hline l_6 & W_3 & W_4 & W_5 \end{array}$$

where

W_3 = "consultation with a specialist is needed";

W_4 = "consultation with a neurologist is needed";

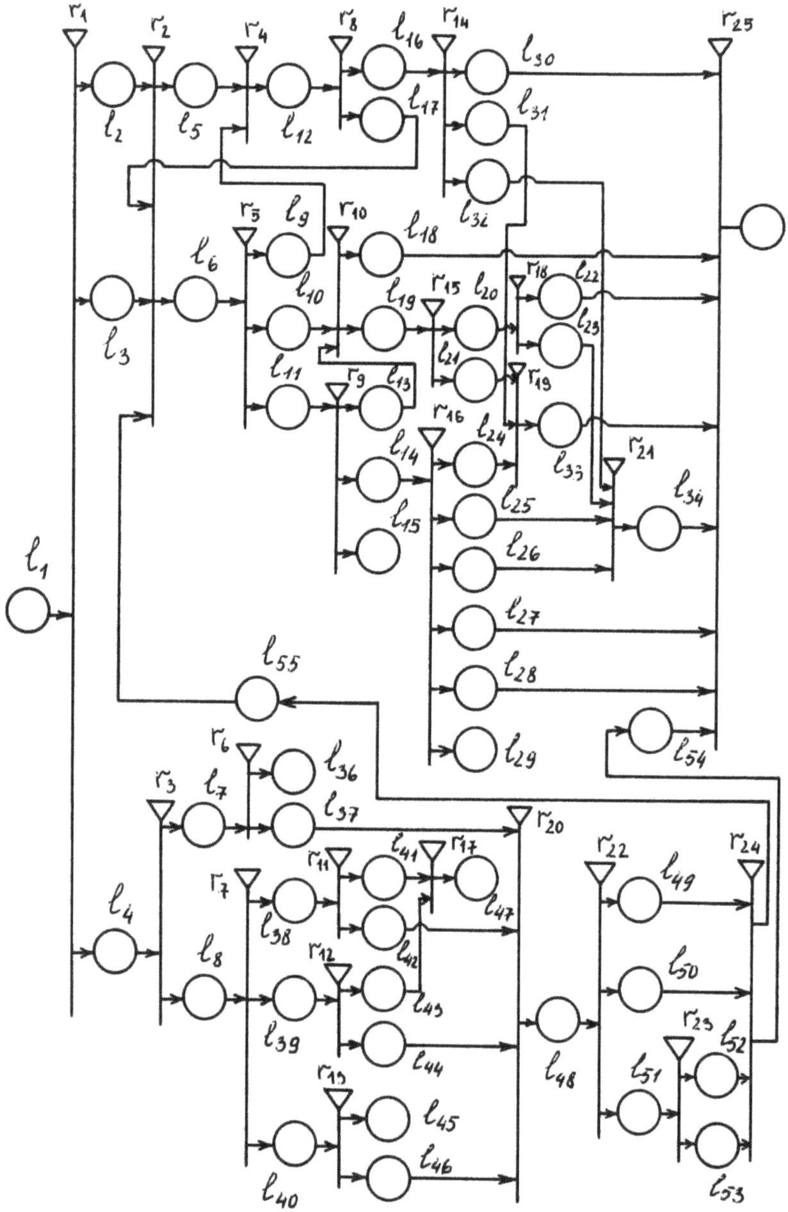
r1
r2
r4
r8
r14
r25
l2
l5
l12
l16
l17
l30
l31
l32
l35
r5
l9
r10
l18
l6
l10
l19
r15
l20
r18
l22
l23
l3
l21
l11
r9
l13
r19
r16
l24
l33
r21
l14
l25
l34
l15
l26
l1
l27
l28
l55
l29
l54
r6
l36
r3
l7
l37
r20
r11
l41
r17
r22
r24
r7
l47
l49
l4
l38
l42
r12
l8
l43
l50
l39
l44
l48
r23
l52
l51
r13
l45
l46
l40
l53

W_5 = "urodynamic investigation is needed";

$$r_4 = \begin{array}{c|c} & l_{12} \\ \hline l_5 & true \\ l_9 & true \end{array} .$$

From places l_5 and l_9 the tokens get into place l_{12} with the following characteristic: "a specialist's opinion is required" (e.g. urologist, gynecologist, etc.). From there the token gets back to l_6, passing through place l_{17} or to l_{16} according to the condition

$$r_8 = \begin{array}{c|cc} & l_{16} & l_{17} \\ \hline l_{12} & W_6 & W_7 \end{array}$$

where

W_6 = "a definite cause of dysuria is found",

$W_7 = \neg W_6$.

The transition condition

$$r_9 = \begin{array}{c|ccc} & l_{13} & l_{14} & l_{15} \\ \hline l_{11} & W_8 & W_9 & W_{10} \end{array}$$

where

W_8 = "the urodynamic tests are normal";

W_9 = "the urodynamic tests are pathologic";

W_{10} = "the urodynamic tests are inconclusive".

When tokens reach place l_{15}, they leave the net with the characteristic "further medical observation is needed".

The transition condition

$$r_{10} = \begin{array}{c|cc} & l_{18} & l_{19} \\ \hline l_{10} & W_{11} & W_{12} \\ l_{13} & false & true \end{array}$$

where

W_{11} = "a neurological condition that explains the micturition disorders is present";

$W_{12} = \neg W_{11}$.

From place l_{18} (through place l_{35}), the token leaves the net with the final characteristic "the micturition disorder is symptomatic of a neurological condition". Depending on the need of psychiatric consultation (the transition r_{15} has a form:

$$r_{15} = \begin{array}{c|cc} & l_{20} & l_{21} \\ \hline l_{19} & W_{13} & W_{14} \end{array}$$

where

W_{13} = "the patient must be consulted by a psychiatrist";

$W_{14} = \neg W_{13}$). The tokens get into place l_{20} or l_{21}. From place l_{20}, the token gets into place l_{22} or l_{23} (the transition r_{18} has a form

$$r_{18} = \begin{array}{c|cc} & l_{22} & l_{23} \\ \hline l_{20} & W_{15} & W_{16} \end{array}$$

where

W_{15} = "the micturition disorder is of functional or anorganic origin";

$W_{16} = \neg W_{15}$).

The transition condition

$$r_{16} = \begin{array}{c|cccccc} & l_{24} & l_{25} & l_{26} & l_{27} & l_{28} & l_{29} \\ \hline l_{14} & W_{17} & W_{18} & W_{19} & W_{20} & W_{21} & W_{22} \end{array}$$

where

W_{17} = "vesical hypercontractility is present";

W_{18} = "a clearcut pathologic vesico-urethral dysfunction is manifested";

W_{19} = "vesico-sphincteric dyssynergism is present";

W_{20} = "cervical obstruction is detected";

W_{21} = "there is vesical hypocontractility";

W_{22} = " other findings are present".

The transition conition

$$r_{14} = \begin{array}{c|ccc} & l_{30} & l_{31} & l_{32} \\ \hline l_{16} & W_{23} & W_{24} & W_{25} \end{array}$$

where

W_{23} = "there is a definite diagnosis";

W_{24} = "there are uro-gynecological disorders";

$W_{25} = \neg W_{24}$.

Thus, from place l_{16} the token gets into places l_{31}, or l_{32}, or l_{30} depending on the characteristics it has acquired in place l_{12} (there is or is not uro-gynecological disorder, or another disease is present).

When a token splits and engenders tokens in places l_2 and l_3, and when at a given moment these newly engendered tokens get into places l_{31}, l_{21} and l_{24} (i.e. there is an accumulation of data for a given patient from different channels), these three tokens fuse into one in place l_{33} with the characteristic "micturition disorders of urogynecological origin", because of the transition condition

$$r_{19} = \begin{array}{c|c} & l_{33} \\ \hline l_{21} & true \\ l_{31} & true \\ l_{24} & true \end{array} .$$

The transition condition

$$r_{21} = \begin{array}{c|c} & l_{34} \\ \hline l_{32} & true \\ l_{23} & true \\ l_{24} & true \\ l_{25} & true \end{array} .$$

Therefore, when tokens engendered by an initial one simultaneously get into places l_{32}, l_{23}, l_{25} and l_{26}, they fuse into one token in place l_{34} with the characteristic "micturition disorders of the isolated neurogenic bladder".

From l_{29} the token leaves the net with the characteristic "non-dysuric disease".

The transition condition

$$r_{25} = \begin{array}{c|c} & l_{35} \\ \hline l_{30} & true \\ l_{18} & true \\ l_{22} & true \\ l_{33} & true \\ l_{34} & true \\ l_{27} & true \\ l_{28} & true \\ l_{54} & true \end{array}.$$

Therefore, from each of the places l_{30}, l_{18}, l_{22}, l_{33}, l_{34}, l_{27}, l_{28} and l_{54}, the token gets into l_{35} which symbolizes the performance of an appropriate treatment.

The transition condition

$$r_3 = \begin{array}{c|cc} & l_7 & l_8 \\ \hline l_4 & true & true \end{array}.$$

The token from l_4 splits into two new tokens which get into l_7 (investigation for hyperazotemia) and l_8 (urinalysis).

The transition condition

$$r_6 = \begin{array}{c|cc} & l_{36} & l_{37} \\ \hline l_7 & W_{26} & W_{27} \end{array}$$

where

W_{26} = "elevated blood urea nitrogen is present";

$W_{27} = \neg W_{26}$

From l_7 the token gets into place l_{36} or l_{37} according to condition r_7 (hyperazotemia is present or is not). In the first case the token leaves the net and gets into the GN from § 4.1.4.

From l_8, the token splits into three which get correspondingly to places: l_{38} ("two glasses" test), l_{39} (bacteriologic investigations, including for b.Koch) and l_{40} (micturition flow), because

$$r_7 = \begin{array}{c|ccc} & l_{38} & l_{39} & l_{40} \\ \hline l_7 & true & true & true \end{array}.$$

The transition conditions r_{11}, r_{12} and r_{13} have the form

$$r_{11} = \begin{array}{c|cc} & l_{41} & l_{42} \\ \hline l_{38} & W_{28} & W_{29} \end{array}$$

where

W_{28} = "turbid urine is found";

$W_{29} = \neg W_{28}$;

$$r_{12} = \begin{array}{c|cc} & l_{43} & l_{44} \\ \hline l_{39} & W_{30} & W_{31} \end{array}$$

where

W_{30} = "a positive bacteriologic finding is present";

$W_{31} = \neg W_{30}$;

r_{13} =	l_{45}	l_{46}
l_{40}	W_{32}	W_{33}

where

W_{32} = "there is an increased urine flow";

W_{33} = $\neg W_{32}$.

The increase of urine flow means polyuria, so the token in place l_{45} leaves the net with this characteristic.

From places l_{37}, l_{42}, l_{44} and l_{46} the tokens get into place l_{48} (combined excretory and mictional urography) because

r_{20} =	l_{48}
l_{37}	true
l_{42}	true
l_{44}	true
l_{46}	true

The transition conditions r_{22}, r_{23}, r_{24} and r_{25} have the form

r_{22} =	l_{49}	l_{50}	l_{51}
l_{48}	W_{34}	W_{35}	W_{36}

where

W_{34} = "the X-ray picture reveals distinct pathological finding",

W_{35} = $\neg W_{34}$,

W_{36} = "the X-ray report is in conclusive".

$$r_{23} = \begin{array}{c|cc} & l_{52} & l_{53} \\ \hline l_{51} & W_{37} & W_{38} \end{array}$$

where

W_{37} = "suprapubic cysto-urethrography is performed";

W_{38} = "retrograde urethrography is performed" ;

$$r_{24} = \begin{array}{c|cc} & l_{54} & l_{55} \\ \hline l_{49} & W_{39} & W_{40} \\ l_{50} & W_{39} & W_{40} \\ l_{52} & W_{39} & W_{40} \\ l_{53} & W_{39} & W_{40} \end{array}$$

where

W_{39} = "there is a definitive diagnosis";

$W_{40} = \neg W_{39}$.

Finally,

$$r_{17} = \begin{array}{c|c} & l_{47} \\ \hline l_{41} & \text{true} \\ l_{43} & \text{true} \end{array}$$

* * *

§ 4.1.7 *Arterial hypertension*

The purpose of the diagnostic examination of a patient is with hypertension to determine both the severity of the disease and whether the hypertension is primary (essential) or secondary to another disorder. The results of this procedure will determine the need for treatment as well as the type of treatment.

The systemic arterial hypertension is usually defined as a diastolic pressure (DBP) of 90 mmHg or higher recorded on several examinations of a subject with normal salt intake (10 ± 3 g/day) regardless of age and sex.

A token in place l_1 means a patient with arterial hypertension (as initial characteristic).

The transition conditions of the GN have the form

$$r_1 = \begin{array}{c|ccccc} & l_2 & l_3 & l_4 & l_5 & l_6 \\ \hline l_1 & W_1 & W_2 & W_3 & W_4 & W_5 \end{array}$$

where

W_1 = "recording the DBP under appropriate conditions";

W_2 = "fundoscopic grading of retinal changes (according Keith-Wagner-Barker classification)";

W_3 = "each of left ventricular hypertrophy";

W_4 = "investigation for renal damage";

r_5 = "evidence of target organ damage or presence of cardiovascular risk factor other than hypertension";

$$r_2 = \begin{array}{c|ccc} & l_7 & l_8 & l_9 \\ \hline l_2 & W_6 & W_7 & W_8 \end{array}$$

where

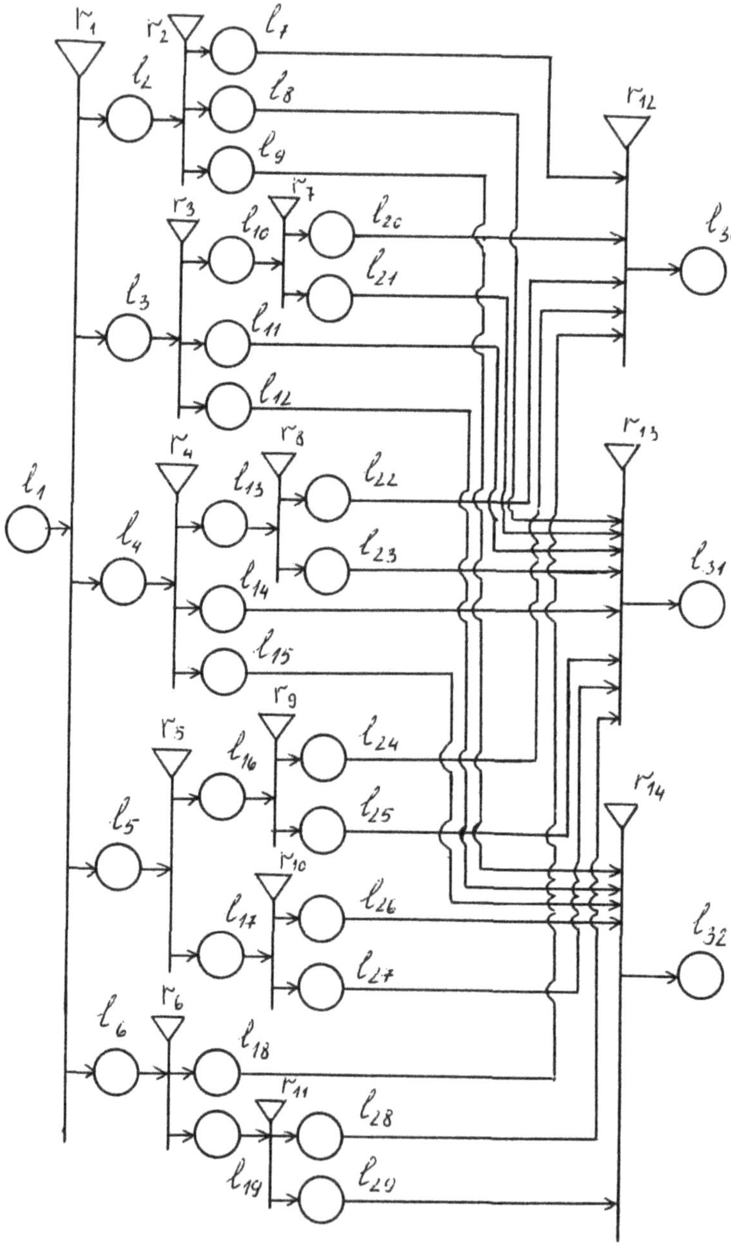

r1
r2
l7
l2
l8
r12
l9
r7
r3
l10
l20
l30
l21
l3
l11
l12
r13
r8
r4
l13
l22
l1
l4
l23
l31
l14
l15
r9
r5
l16
l24
r14
l5
l25
r10
l26
l17
l32
l27
l6
r6
l18
r11
l28
l19
l29

W_6 = "the DBP is within the limits 90-104 mmHg";

W_7 = "the DBP is within the limits 105-114 mmHg";

W_8 = "the DBP is $\geq$ 115 mmHg";

$$r_3 = \begin{array}{c|ccc} & l_{10} & l_{11} & l_{12} \\ \hline l_3 & W_9 & W_{10} & W_{11} \end{array}$$

where

W_9 = "no retinal changes or such of KWB group 1";

W_{10} = "retinal changes KWB group 2";

W_{11} = "retinal changes KWB groups 3 and 4";

$$r_4 = \begin{array}{c|ccc} & l_{13} & l_{14} & l_{15} \\ \hline l_4 & W_{12} & W_{13} & W_{14} \end{array}$$

where

$W_{12} = \neg W_{13} \;\&\; \neg W_{14}$;

W_{13} = "paterns of left ventricular 'strain'";

W_{14} = "left ventricular hypertrophy is present";

$$r_5 = \begin{array}{c|cc} & l_{16} & l_{17} \\ \hline l_5 & W_{15} & W_{16} \end{array}$$

where

W_{15} = "no renal damage";

$W_{16} = \neg W_{15}$;

$$r_6 = \begin{array}{c|cc} & l_{18} & l_{19} \\ \hline l_6 & W_{17} & W_{18} \end{array}$$

where

W_{17} = "no evidence of other target organ damage or other cardiovascular risk factors";

$W_{18} = \neg W_{17}$;

$$r_7 = \frac{l_{20} \quad l_{21}}{l_{10} \mid W_{19} \quad W_{20}}$$

where

W_{20} = "there is a token, related to the present token, at least in one of the following places: l_8, l_9, l_{13}, l_{14}, ..., l_{17}, l_{19}";

$W_{19} = \neg W_{20}$;

$$r_8 = \frac{l_{22} \quad l_{23}}{l_{13} \mid W_{21} \quad W_{22}}$$

where

W_{22} = "there is a token, related to the present token, at least in one of the following places: l_8, l_9, l_{11}, l_{12}, l_{17}, l_{19}";

$W_{21} = \neg W_{22}$;

$$r_9 = \frac{l_{22} \quad l_{25}}{l_{16} \mid W_{23} \quad W_{24}}$$

where

W_{24} = "there is a token, related to the present token, at least in one of the following places: l_8, l_9, l_{13}, l_{14}, l_{15}, l_{19}";

$W_{23} = \neg W_{24}$;

$$r_{10} = \frac{l_{26} \quad l_{27}}{l_{17} \mid W_{25} \quad W_{26}}$$

where

W_{26} = "there is a token, related to the present token, at least in one of the following places: l_9, l_{12}, l_{16}, l_{19}";

$W_{25} = \neg W_{26}$;

$$r_{11} = \begin{array}{c|cc} & l_{28} & l_{29} \\ \hline l_{19} & W_{27} & W_{28} \end{array}$$

where

W_{28} = "there is a token, related to the present token, at least in one of the following places: l_9, l_{12}, l_{15}, l_{17}";

$W_{27} = \neg W_{28}$.

The remaining transition conditions' predicates (of the 12-th, 13-th and 14-th transitions) are only "true".

*

* *

§ 4.1.8 Arterial hypertension of renal origin

The tokens get into place l_1 with initial characteristics formed by three groups of data:

- group X: "signs and symptoms directing to renal origin of the arterial hypertension"
- group Y: "signs and symptoms directing to renovascular origin of the arterial hypertension"
- group Z: "there are no signs, symptoms or conditions, excluding the primary participation of kidneys in the origin of the hypertensive state".

The transition condition

$$r_1 = \begin{array}{c|cc} & l_2 & l_3 \\ \hline l_1 & true & true \end{array}.$$

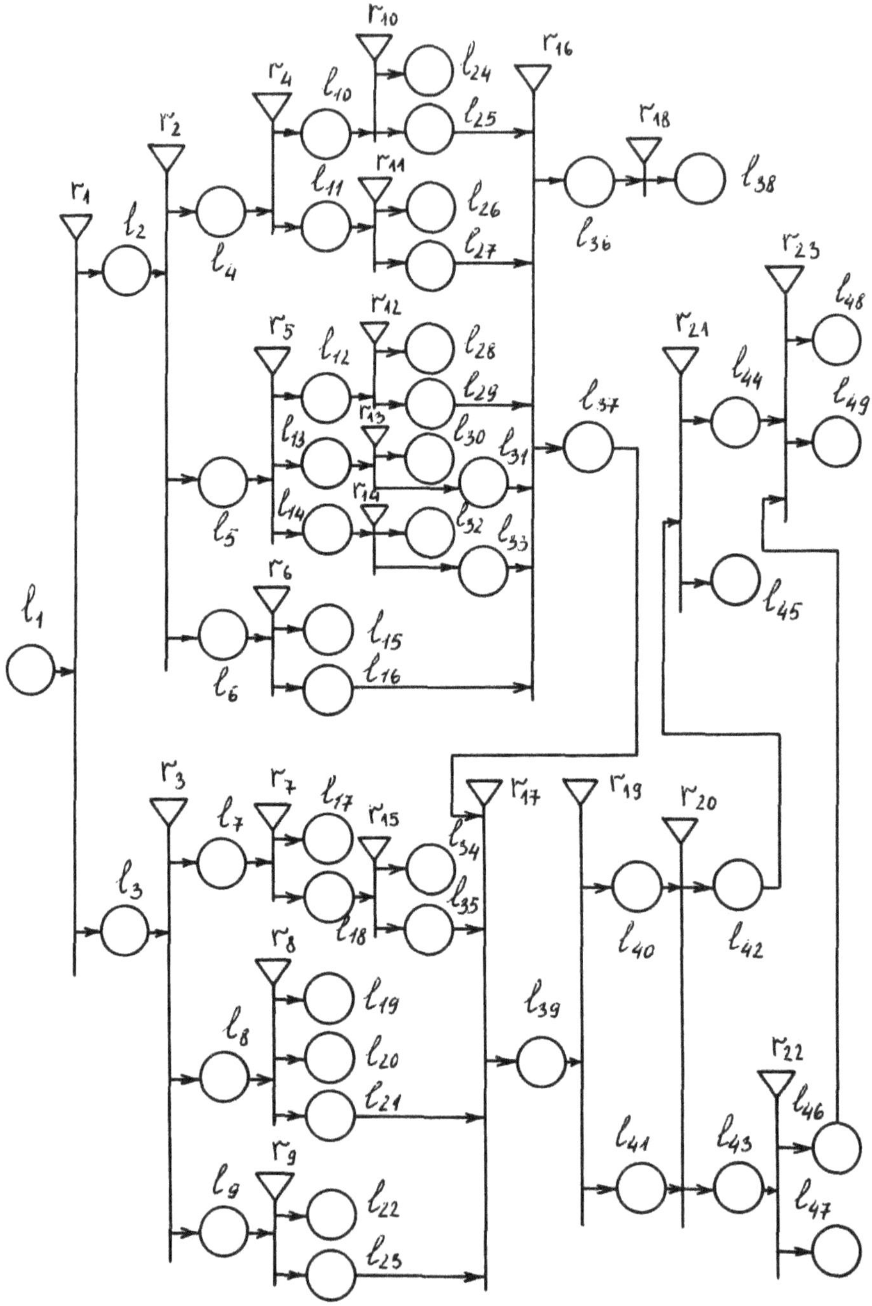
r1
r2
r3
r4
r5
r6
r7
r8
r9
r10
r11
r12
r13
r14
r15
r16
r17
r18
r19
r20
r21
r22
r23
l1
l2
l3
l4
l5
l6
l7
l8
l9
l10
l11
l12
l13
l14
l15
l16
l17
l18
l19
l20
l21
l22
l23
l24
l25
l26
l27
l28
l29
l30
l31
l32
l33
l34
l35
l36
l37
l38
l39
l40
l41
l42
l43
l44
l45
l46
l47
l48
l49

Thus the token splits into two, which get into place l_2 (urine is sent for analysis) and place l_3 (blood sample is taken for examination). The token in place l_2 splits into three new tokens that simultaneously get into places l_4 (biochemical examination of urine), l_5 (urinary sediment examination) and l_6 (bacteriologic examination) because transition condition r_2 has the form

$$r_2 = \begin{array}{c|ccc} & l_4 & l_5 & l_6 \\ \hline l_2 & true & true & true \end{array}.$$

The transition conditions r_4 and r_5 have similar form

$$r_4 = \begin{array}{c|cc} & l_{10} & l_{11} \\ \hline l_4 & true & true \end{array};$$

$$r_5 = \begin{array}{c|ccc} & l_{12} & l_{13} & l_{14} \\ \hline l_5 & true & true & true \end{array}.$$

When the token gets into places $l_{10}, \ldots, l_{14}$, if means testing simultaneously for proteinuria, glucosuria, hematuria, leucocyturia and bacteriuria.

The transition conditions r_6, r_{10}, $r_{11}, \ldots, r_{14}$ have the form

$$r_6 = \begin{array}{c|cc} & l_{15} & l_{16} \\ \hline l_6 & W_1 & W_2 \end{array}$$

where

W_1 = "there is positive urine culture",

$W_2 = \neg W_1$;

$$r_{10} = l_{10} \mid \frac{l_{24} \qquad l_{25}}{W_3 \qquad W_4}$$

where

W_3 = "there is protenuria";

$W_4 = \neg W_3$;

$$r_{11} = l_{10} \mid \frac{l_{26} \qquad l_{27}}{W_5 \qquad W_6}$$

where

W_5 = "diabetes mellitus is suspected";

$W_6 = \neg W_5$;

$$r_{12} = l_{12} \mid \frac{l_{28} \qquad l_{29}}{W_7 \qquad W_8}$$

where

W_7 = "there is hematuria";

$W_8 = \neg W_7$;

$$r_{13} = l_{13} \mid \frac{l_{30} \qquad l_{31}}{W_9 \qquad W_{10}}$$

where

W_9 = "there are bacteria in urine sediment";

$W_{10} = \neg W_9$;

$$r_{14} = \begin{array}{c|cc} & l_{32} & l_{33} \\ \hline l_{14} & W_{11} & W_{12} \end{array}$$

where

W_{11} = "there is bacteriuria";

$W_{12} = \neg W_{11}$;

The transition condition

$$r_{17} = \begin{array}{c|cc} & l_{36} & l_{37} \\ \hline l_{25} & W_{13} & W_{14} \\ l_{27} & W_{13} & W_{14} \\ l_{29} & W_{13} & W_{14} \\ l_{31} & W_{13} & W_{14} \\ l_{33} & W_{13} & W_{14} \\ l_{16} & W_{13} & W_{14} \end{array}$$

where

W_{13} = "glomerular nephropathy is suspected";

$W_{14} = \neg W_{13}$.

The token in place l_3 splits into three tokens which get into places l_7 (serum creatinine by urea determination), l_8 (serum potassium), and l_9 (blood sugar), because the transition condition r_3 has the form:

$$r_3 = \begin{array}{c|ccc} & l_7 & l_8 & l_9 \\ \hline l_3 & true & true & true \end{array} .$$

The transition conditions r_7, r_8, r_9, r_{10} and r_{17} have the form

$$r_7 = \begin{array}{c|cc} & l_{17} & l_{18} \\ \hline l_7 & W_{15} & W_{16} \end{array}$$

where

W_{15} = "there is increase in serum creatinine level";

$W_{16} = \neg W_{15}$;

$$r_8 = \begin{array}{c|ccc} & l_{19} & l_{20} & l_{21} \\ \hline l_8 & W_{17} & W_{18} & W_{19} \end{array}$$

where

W_{17} = "hyperpotassemia is present";

W_{18} = "normopotassemia is present";

W_{19} = "hypopotassemia is present";

$$r_9 = \begin{array}{c|cc} & l_{22} & l_{23} \\ \hline l_9 & W_{20} & W_{21} \end{array}$$

where

W_{20} = "there is hyperglycemia";

$W_{21} = \neg W_{20}$;

$$r_{10} = \begin{array}{c|cc} & l_{34} & l_{35} \\ \hline l_{18} & W_{22} & W_{23} \end{array}$$

where

W_{22} = "there is a low glomerular filtration rate";

$W_{23} = \neg W_{22}$;

	l_{39}
r_{17} = l_{37}	true
l_{35}	true
l_{21}	true
l_{23}	true

All tokens pertaining to a specific patient when getting into place l_{39} fuse into one token with characteristics incorporating all the previously acquired characteristics by the tokens resulting from the investigations carried out up to now. Place l_{39} symbolizes "a test for iodine sensitivity is performed".

The transition conditions r_{19} and r_{20} have the form

	l_{40}	l_{41}
r_{19} = l_{39}	W_{24}	W_{25}

where

W_{24} = "the test for sensitivity is negative";

W_{25} = $\neg W_{24}$;

	l_{42}	l_{43}
r_{20} = l_{40}	W_{26}	W_{27}
l_{41}	false	true

where

W_{26} = "intravenous urography is performed";

W_{27} = "other instrumental investigations of kidneys are necessary (isotopic, sonography, etc.)".

The place l_{43} corresponds to the following actions: isotopic investigations (renography, scintigraphy, gamma-camera), sonography, computer tomography, and the place l_{42} to excretory urography.

The transition conditions r_{21}, r_{22} and r_{23} have the form

$$r_{21} = \begin{array}{c|cc} & l_{44} & l_{45} \\ \hline l_{42} & W_{28} & W_{29} \end{array}$$

where

$W_{28} = \neg W_{29}$;

W_{29} = "there are urographic data for renovascular hypertension";

$$r_{22} = \begin{array}{c|cc} & l_{46} & l_{47} \\ \hline l_{43} & W_{30} & W_{31} \end{array}$$

where

$W_{30} = \neg W_{31}$;

W_{32} = "there are signs of renovascular hypertension";

$$r_{23} = \begin{array}{c|cc} & l_{48} & l_{49} \\ \hline l_{44} & W_{32} & W_{33} \\ & & \\ l_{46} & W_{32} & W_{33} \end{array}$$

where

$W_{32} = \neg W_{33}$;

W_{33} = "the initial characteristics of the token has positive signs from group Y".

When tokens get into place l_{24} there is a call to GN from

§ 4.1.3, in place l_{28} - to GNs from § 4.1.2 and § 4.1.5, in places l_{15} and l_{30} to GNs from § 4.1.3 and § 4.1.5; in places l_{17}, l_{19} and l_{34} to GN from § 4.1.4; in place l_{32} - to GN from § 4.1.3; in places l_{47} and l_{49} - to GN for § 4.1.9.

The tokens in places l_{20}, l_{22} and l_{26} leave the GN because of the indications for other diseases, and in places l_{38} and l_{48} they receive the characteristic: appropriate treatent and medical observation of the patient.

*
* *

§ 4.1.9 Renovascular hypertension

The tokens (patients) get into place l_1 from a GN from § 4.1.8 with the initial characteristics "signs and symptoms directing to renovascular origin of the arterial hypertension".

The transition condition r_1 has the form:

$$r_1 = \begin{array}{c|cccc} & l_2 & l_3 & l_4 & l_5 \\ \hline l_1 & true & true & true & true \end{array}.$$

A token from place l_1 splits into four tokens that get into places $l_2, \ldots, l_5$. These places signify: l_2 - check-up for endocrine disorder (primary hyperaldosteronism, pheochromocytoma, hypercorticism etc.), l_3 - performing an iodine sensitivity test, l_4 - investigation of plasma renin activity, l_5 - performance of pharmacological tests.

The transition condition

$r_2 =$

	l_6	l_7
l_2	W_1	W_2

where

W_1 = "an endocrine disease as a probable cause of the hypertensive state is found",

$W_2 = \neg W_1$.

Place l_6 is the final one for the net.

The transition conditions r_3, r_4, r_5, r_8, r_9 and r_6 have the forms

$r_3 =$

	l_8	l_9
l_3	W_3	W_4
l_{10}	W_3	W_4
l_{17}	W_3	W_4
l_{19}	W_3	W_4

where

W_3 = "pharmacological test gives negative result and it is possible to perform a vasography";

$W_4 = \neg W_3$;

W_5 = "high level of plasma renin activity is found";

$W_6 = \neg W_5$;

$r_5 =$

	l_{12}	l_{13}
l_5	W_7	W_8

where

W_7 = "angiotensin test is performed";

W_8 = "test with saralasin is performed";

$$r_8 = \begin{array}{c|cc} & l_{17} & l_{18} \\ \hline l_{12} & W_9 & W_{10} \end{array}$$

where

W_9 = "the angiotensin test is positive";

$W_{10} = \neg W_9$;

$$r_9 = \begin{array}{c|cc} & l_{19} & l_{20} \\ \hline l_{13} & W_{11} & W_{12} \end{array}$$

where

W_{11} = "the saralasin test is positive";

$W_{12} = \neg W_{11}$;

$$r_6 = \begin{array}{c|c} & l_{14} \\ \hline l_7 & true \\ & \\ l_8 & true \end{array}$$

The tokens from places l_7 and l_8 reach directly place l_{14} (performance of vasography - abdominal aortography and/or selective renal arteriography).

The transition conditions r_7, r_{10}, r_{11} and r_{12} have the form

$$r_{10} = \begin{array}{c|cc} & l_{21} & l_{22} \\ \hline l_{14} & W_{15} & W_{16} \end{array}$$

where

W_{15} = "there are signs of renal arterial stenosis";

$W_{16} = \neg W_{15}$;

$$r_{11} = \begin{array}{c|cc} & l_{23} & l_{24} \\ \hline l_{21} & W_{17} & W_{18} \\ l_{15} & \text{true} & \text{false} \end{array}$$

where

W_{17} = "there is evidence that the stenosis is functionally significant",

$W_{18} = \neg W_{17}$;

$$r_{12} = \begin{array}{c|ccccc} & l_{25} & l_{26} & l_{27} & l_{28} & l_{29} \\ \hline l_{23} & \text{true} & \text{true} & \text{true} & \text{true} & \text{true} \end{array}.$$

Places l_{25}, l_{26}, ..., l_{29} indicate the performance of appropriate tests to determine whether the stenosis has functional significance: l_{25} (Howard's test), l_{26} (Rapoport or Stamey test), l_{27} (separate hemodynamic investigation of kidneys), l_{28} (determination of plasma renin activity in renal veins separately), l_{29} (other investigations of separate kidney functions).

$$r_{13} = \begin{array}{c|cc} & l_{30} & l_{31} \\ \hline l_{25} & W_{19} & W_{20} \end{array}$$

where

W_{19} = "the test of Howard is positive",

$W_{20} = \neg W_{19}$;

$$r_{14} = \begin{array}{c|cc} & l_{32} & l_{33} \\ \hline l_{26} & W_{21} & W_{22} \end{array}$$

where

W_{21} = "the test of Rapoport or Stamey is positive";

$W_{22} = \neg W_{21}$;

$$r_{15} = \begin{array}{c|cc} & l_{34} & l_{35} \\ \hline l_{27} & W_{23} & W_{24} \end{array}$$

where

W_{23} = "unilateral decrease of ERBF is found during separate hemodynamic isotopic investigation of Kidneys";

$W_{24} = \neg W_{23}$;

$$r_{16} = \begin{array}{c|cc} & l_{36} & l_{37} \\ \hline l_{28} & W_{25} & W_{26} \end{array}$$

where

W_{25} = "augmented plasma renin activity is determined unilaterally";

$W_{26} = \neg W_{25}$;

$$r_{17} = \begin{array}{c|cc} & l_{38} & l_{39} \\ \hline l_{29} & W_{27} & W_{28} \end{array}$$

where

W_{27} = "other functional tests for determining unilateral renal lesion";

$W_{28} = \neg W_{27}$.

$$r_{18} = \begin{array}{c|cc} & l_{40} & l_{41} \\ \hline l_{30} & W_{29} & W_{30} \\ l_{32} & W_{29} & W_{30} \\ l_{34} & W_{29} & W_{30} \\ l_{36} & W_{29} & W_{30} \\ l_{38} & W_{29} & W_{30} \end{array}$$

where

W_{29} = "a biopsy of the contralateral kidney is necessary";

$W_{30} = \neg W_{29}$.

Place l_{40} signifies the performance of renal biopsy.

$$r_{19} = \begin{array}{c|cc} & l_{42} & l_{43} \\ \hline l_{40} & W_{31} & W_{32} \end{array}$$

where

W_{31} = "evidence for lesion of the contralateral kidney is present";

$W_{32} = \neg W_{31}$.

$$r_{20} = \begin{array}{c|c} & l_{44} \\ \hline l_{43} & true \\ l_{41} & true \end{array}$$

Place l_{44} signifies aa check-up for contraindications to surgical intervention. The transition condition

$$r_{21} = \begin{array}{c|cc} & l_{45} & l_{46} \\ \hline l_{44} & W_{33} & W_{34} \end{array}$$

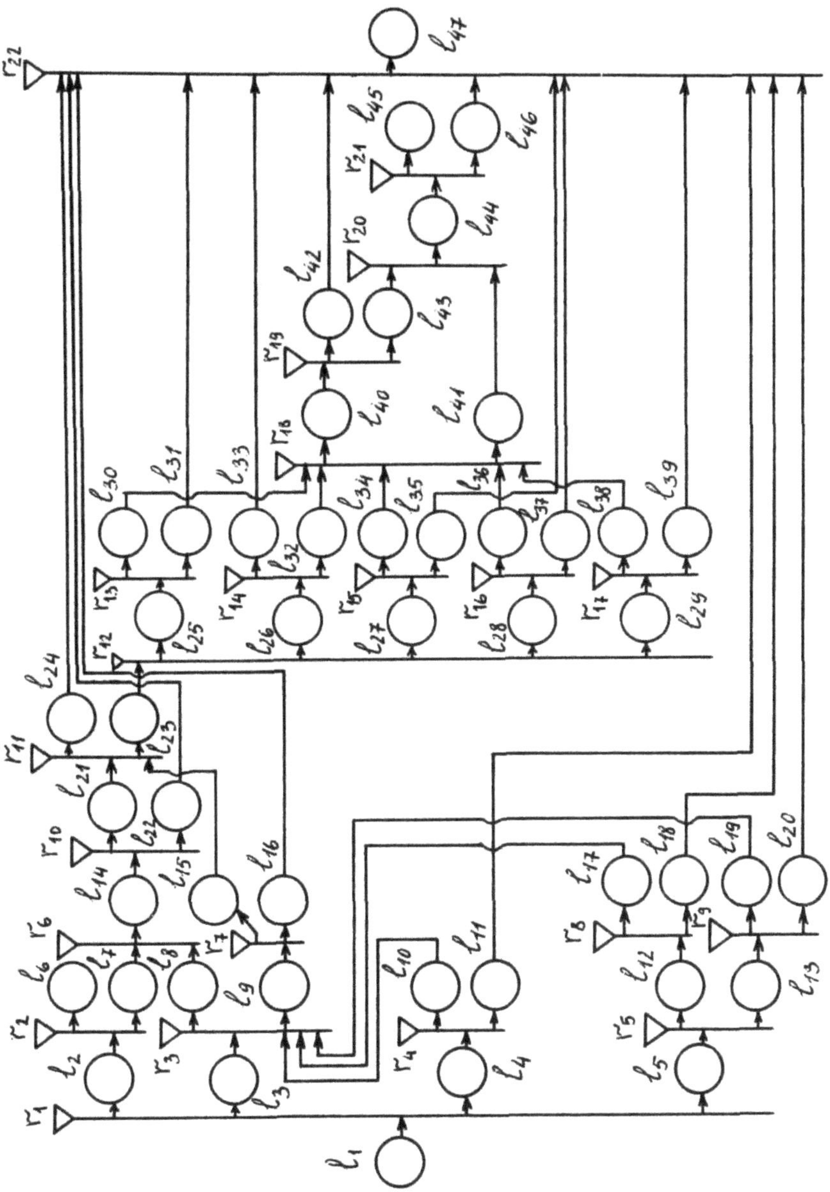
l1
l2
l3
l4
l5
l6
l7
l8
l9
l10
l11
l12
l13
l14
l15
l16
l17
l18
l19
l20
l21
l22
l23
l24
l25
l26
l27
l28
l29
l30
l31
l32
l33
l34
l35
l36
l37
l38
l39
l40
l41
l42
l43
l44
l45
l46
l47
r1
r2
r3
r4
r5
r6
r7
r8
r9
r10
r11
r12
r13
r14
r15
r16
r17
r18
r19
r20
r21
r22

where

W_{33} = "there are contraindications";

$W_{34} = \neg W_{33}$

(a surgical investigation (l_{45}) is possible).

From places l_{11}, l_{16}, l_{18}, l_{20}, l_{22}, l_{24}, l_{31}, l_{33}, l_{35}, l_{37}, l_{39}, l_{42} and l_{46} the tokens get into place l_{47}. Here, they acquire the characteristic "appropriate treatment and medical observation of the patient".

REFERENCES:

[1] F. Balag, G. Petranyi and F. Renyi-Vamos, Nephrologia, Medicina Konyvkiado, Budapest, 1980 (in Hung.).

[2] B. Brenner and J. Lazarus (Eds.), Acute renal failure, Saunders Co., Philadelphia, 1983.

[3] B. Brenner and F. Rector (Eds.), The kidney, Vol. 1 and 2, Saunders Co., Philadelphia, 1986.

[4] N. Bricker and M. Kirschenbaum (Eds.), The kidney - diagnosis and menagement, Wiley Medical Publ., New York, 1984.

[5] J. A. Halsted and C. H. Halsted (Eds.) The laboratory in clinical medicine - interpretation and application, Saunders Co., Philadelphia, 1981.

[6] J. Hamburger, J. Crosnier and J.-P. Grunfeld (Eds.), Nephrologie, Vol. 1 and 2, Flammarion, Paris, 1979 (in French).

[7] J. Jones, J. Briggs and T. Hargreare, Diagnosis and menagement of renal and urinary diseases, Blackwell, Oxford, 1982.

[8] S. Klahr and S. Massry, Contemporary Nephrology, Plenum Publ. Co. New York, 1981.

[9] H. Kremling, W. Lutzeyer and R. Heintz, Gynakologische Urologie

und Nephrologie, Urban Schwarzenberg, Munchen, 1982 (in German).

[10] M. Legran and J.-M. Suc, Abrege de nephrologie, Masson, Paris, 1981 (in French).

[11] G. Majdrakov and . Popov (Eds.) Gidney diseases, Sofia, Medicina i Fizkultura, 1980 (in Bulg.).

[12] F. Marsh (Ed.) Postgraduate nephrology, W. Heinemann Medical Books Ltd., London, 1985.

[13] S. Massry and R. Glassock, Textbook of nephrology, Vol. 1 and 2, Williams and Wilkins, Baltimore, 1983.

[14] E. Patev, Acute renal failure, Sofia, Medicina i Fizkultura, 1981 (in Bulg.).

[15] D. Vasilev, Vasorenal hypertension, Sofia, Medicina i Fizkultura, 1989 (in Bulg.).

[16] D. Vasilev and G. Stefanov (Eds.), Clinical nephrology, Sofia, Medicina i Fizkultura, 1990 (in Bulg.).

[17] J. Sorsich and K. Atanassov. Application of generalized nets in medicine (Acute Attack of Gouty Arthritis), Proc. of Third Symp. Int. Ing. Biomed., Madrid, Oct. 1987, 643-645.

[18] R. Bellman, Mathematical methods in medicine. World Sci. Publ., 1983.

[19] S. Kiforenko, Methods of mathematical biology, Kiev, Vischa skola, 1984 (in Russian).

[20] J. Piemann, Modeling hospital information systems with Petri nets Methods Inf. Medicine, Vol. 27, No. 1, 1988, 17-22.

[21] Mathematical methods in medicine and biology, Proc. of UNC of Academy of Sciences of USSR, Sverdlovsk, 1986 (in Russian).

[22] J. Sorsich, An example of application of generalized nets in medicine. Proc of II Int Symp. "Automation and Scientific Instrumentation", Varna, May 1983, 387-389.

[23] J. Sorsich and K. Atanassov, Application of generalized nets in medicine (Diagnostics of arterial hypertension of renal origin).

Proc. of III Int. School "Automation and Scientific Instrumentation", Varna, 1984, 233-236.

[24] J. Sorsich and K. Atanassov, Application of generalized nets in medicine (Renal colic), Lect. Notes in Medical Inform. 24, 1984, 352-355.

[25] J. Sorsich and K. Atanassov, Application of generalized nets in medicine (Renovascular hypertension). Proc. of Third Int. Symp. "Automation and Sci. Instrumentation", Varna, 1985, 167-169.

[26] J. Sorsich and K. Atanassov, Application of generalized nets in medicine (Permanent Proteinuria), Proc. of National Sci. Session "Automation of Biotechnology Processes", Sofia, Oct. 1985, 138-141 (in Bulgarian).

[27] J. Sorsich and K. Atanassov, Application of generalized nets in medicine (Haematuria), Proc. of III International Symp. "Automation and Scientific Instrumentation", Varna, Oct. 1985, 163-166.

[28] J. Sorsich and K. Atanassov, Application of generalized nets in medicine (Dysuria), Proc. of III Symp. Int. Ing. Biomed., Madrid, Oct. 1987, 639-642.

[29] J. Sorsich, Application of the generalized nets in medicine (Elevated blood used nitrogen), First Sci. Session of the Math. Found. of AI Seminar, Sofia, Oct. 10, 1989, Preprint IM-MFAIS-7-89, Sofia, 1989, 57-59.

[30] J. Sorsich, Application of the generalized nets in medicine (Arterial hypertension), Ninetieth Session of the Nat. Seminar of Informatics of the Union of Bulgarian Mathematicians and Fourth Sci. Session of the Math. Found. AI Seminar, Sofia, Nov. 5, 1990, Preprint IM-MFAIS-5-90, Sofia, 1990, 37-39.

§ 4.2: MODELLING OF DIAGNOSTIC AND THERAPEUTIC PROCESSES IN MEDICINE BY GENERALIZED NETS

Borjana Jordanova and Joseph Sorsich

In this article we describe a general outline of the approach to a patient who needs medical advice.

This net is more general than the ones described in § 4.1. It lacks the details of the others, but it gives some idea of the stages that most patients pass through in a medical department. So the nets described in § 4.1 are its concrete applications.

In the net presented it is possible to see more clearly the ways of introducing waiting and performing times of medical examinations and carrying out the appropriate investigations, as well as the time duration of the whole cycle of treatment. These time-parameters may be obtained by a characteristic function of the net as presented below.

The GN models of the diagnostic process and the appropriate treatment of other (non-nephrological) diseases may also be settled in this scheme.

The token (patient) enters place l_1 with the characteristic "need to consult a physician" (with some complaints or for prophylactic needs).

The GN below described is a reduced GN from the class

$$\Sigma_{A_3, A_4, A_6, A_7, \pi_A, \pi_L, c, \Theta_1, \Theta_2, \pi_K, \Theta_K, b}$$

(everywhere the transition type is "v", the places' and arcs' capacities are "∞", the places' and transitions' priorities are "0", for every token α: $b(\alpha) = \infty$).

The forms of the GN transitions are as follows.

The transition Z_1 is

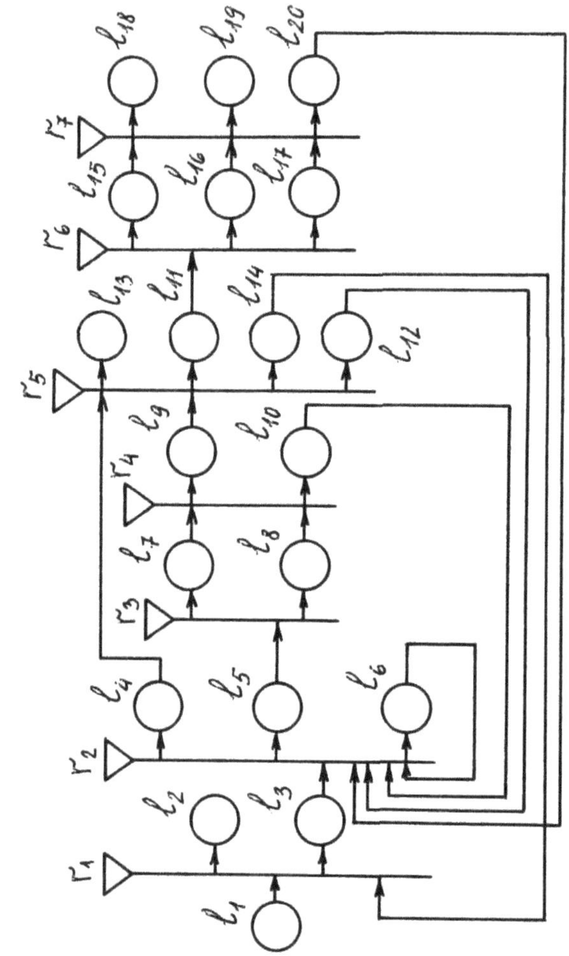

$$Z_1 = \langle \{l_1, l_{14}\}, \{l_2, l_3\}, r_1 \rangle,$$

where

r_1 =

	l_2	l_3
l_1	W_1	W_2
l_{14}	W_1	W_2

where

W_1 = "the result of the medical examination is that the patient is healthy";

$W_2 = \neg W_1$.

The ransition Z_2 is

$$Z_2 = \langle \{l_3, l_6, l_{10}, l_{12}, l_{20}\}, \{l_4, l_5, l_6\}, r_2 \rangle,$$

where

r_2 =

	l_4	l_5	l_6
l_3	W_3	W_4	W_5
l_6	W_3	W_4	W_5
l_{10}	W_3	W_4	W_5
l_{12}	W_3	W_4	W_5
l_{20}	W_3	W_4	W_5

where

$W_3 = \neg W_4 \ \& \ \neg W_5$,

W_4 = "there is need for appropriate investigations";

W_5 = "there is need to consult a specialist".

The transition Z_3 is

$$Z_3 = \langle \{l_5\}, \{l_7, l_8\}, r_3 \rangle,$$

where

$$r_3 = \begin{array}{c|cc} & l_7 & l_8 \\ \hline l_5 & W_6 & W_7 \end{array}$$

where

W_6 = "laboratory investigations are necessary";

W_7 = "instrumental investigations (e.g. X-ray, sonography, ECG etc.) are necessary".

The transition Z_4 is

$$Z_4 = \langle \{l_7, l_8\}, \{l_9, l_{10}\}, r_4 \rangle,$$

where

$$r_4 = \begin{array}{c|cc} & l_9 & l_{10} \\ \hline l_7 & W_8 & W_9 \\ l_8 & W_8 & W_9 \end{array}$$

where

W_8 = "abnormal findings are present";

$W_9 = \neg W_8$.

The transition Z_5 is

$$Z_5 = \langle \{l_4, l_9\}, \{l_{11}, l_{12}, l_{13}, l_{14}\}, r_5 \rangle,$$

where

$$
r_5 = \begin{array}{c|cccc}
 & l_{11} & l_{12} & l_{13} & l_{14} \\
\hline
l_4 & W_{10} & W_{11} & W_{12} & W_{13} \\
l_9 & W_{10} & W_{11} & W_{12} & W_{13}
\end{array}
$$

where

W_{10} = "patient needs appropriate treatment";

W_{11} = "patient needs some complementary investigations";

W_{12} = "patient is estimated to be healthy";

$W_{13} = \neg W_{10} \ \& \ \neg W_{11} \ \& \ \neg W_{12}$.

The transition Z_6 is

$$Z_6 = \langle \{l_{11}\}, \{l_{15}, l_{16}, l_{17}\}, r_6 \rangle,$$

where

$$
r_6 = \begin{array}{c|ccc}
 & l_{15} & l_{16} & l_{17} \\
\hline
l_{11} & W_{14} & W_{15} & W_{16}
\end{array}
$$

where

W_{14} = "ambulatory treatment is suitable";

W_{15} = "treatment at home is recommeanded";

W_{16} = "hospitalization is necessary".

The transition Z_7 is

$$Z_7 = \langle \{l_{15}, l_{16}, l_{17}\}, \{l_{18}, l_{19}, l_{20}\}, r_7 \rangle,$$

where

$$r_7 = \begin{array}{c|ccc} & l_{18} & l_{19} & l_{20} \\ \hline l_{15} & W_{17} & W_{18} & W_{19} \\ l_{16} & W_{17} & W_{18} & W_{19} \\ l_{17} & W_{17} & W_{18} & W_{19} \end{array}$$

where

W_{17} = "patient is restored to health";

W_{18} = "patient is dead";

W_{19} = $\neg W_{17}$ & $\neg W_{18}$.

The tokens receives in the different places the following characteristics by the characteristic function $\Phi = \bigcup_{i=2}^{20}$, defined by:

Φ_2 -> "healthy person";

Φ_3 -> "duration of medical examination";

Φ_4 -> "*";

Φ_5 -> "kinds of appropriate laboratory examinations, the duration to carry out these examinations";

Φ_6 -> "kind(s) of appropriate consultation(s), duration of these consultation(s)";

Φ_7 -> "result(s) of the laboratory investigation(s)";

Φ_8 -> "result(s) of the instrumental investigation(s)";

Φ_9 -> "presence of abnormal findings";

Φ_{10} -> "*";

Φ_{11} -> "patient needs appropriate treatment";

Φ_{12} -> "patient needs complementary investigation(s) or consultation(s)";

Φ_{13} -> "patient is estimated healthy after the consultation(s)";

Φ_{14} -> "further surveillance or control is needed";

Φ_{15} -> "ambulatory treatment";

Φ_{16} -> "treatment at home";

Φ_{17} -> "treatment in hospital";

Φ_{18} -> "the patient is cured with end of treatment, duration of the treatment";

Φ_{19} -> "exitus letalis";

Φ_{20} -> "*".

The characteristic function determines some time-parameters of the model. For their calculation statistical data, as well as results from concrete observations, can be used.

Chapter 5:

GENERALIZED NETS AND COMPUTER SCIENCE

Some applications of the generalized nets as a means for modelling of computers are introduced in § 5.1, § 5.2, ..., and § 5.5. The models are related to the various processes in the computer and they are written by the postgratuate students of the Institute for Microsystems, Sofia.

§ 5.6 deserves special attention where a short description of the Program Package for Generalized Nets (PPGN) is given. This description corresponds to the current state of the work on PPGN and contains the basic directions of its further development.

§ 5.1: IMPLEMENTATION OF GENERALIZED NETS IN DISTRIBUTED OPERATING SYSTEM DEVELOPMENT

Niko Pehlivanov

§ 5.1.1 Introduction

A general view upon the latest developments in operating systems shows that distributed processing is a subject of growing attention. This interest may be explained, on the one hand by the increasing demands for more computing power and, on the other by its enhancements to hardware technology. The advantage of distributed operating systems has now become a fact, after a long period of evolution in computer science, particularly in the field of operating systems. In the last several years that local area networks (LANs) and personal workstations have become so popular and widely used and that serious problems associated with these arise and required solution. Present operating systems provide time-sharing, multiprocessor, graphics, networking, etc. However, the answer to demands for efficient use of shared computing and storage resources is in distributed operating systems.

The design of such distributed systems has proved to be a long and difficult process, which sometimes takes years of hard work. This is caused by the presence of multiple nodes with separate tasks running on them that share system resources and require synchronization, control and load balancing. A major consequence of the latter is the numerous set of states in which the system may appear. Its performance is therefore determined by the ability of the software to act adequately in each particular state, which itself is defined by the respective algorithms.

A general way to facilitate software design is to present the algorithms, the corresponding system states and the transitions between them in a graph scheme. The implementation of Generalized Nets

(GNs) is particularly suitable for this purpose, as they provide a universal method for simulation of system behaviour.

The purpose of this article is to outline some aspects of GN implementation in one specific area of distributed operating system development - the design of distributed file system (DFS).

§ 5.1.2 Generalized Net model for a Distributed File System

The DFS is an element of every distributed operating system, which carries out functions of vital importance. It has to provide the availability and consistency of all files, to control concurrent file access, giving distinct response times to client requests. To meet these requirements, a DFS manages all shared storage resources, placed at the nodes, maintaining all data integrity and security.

One generalized presentation of a DFS is given in Fig. 5.1. It shows several nodes in a distributed system, some of which have storage resources attached to them, while the others do not have such. All system services and hardware components that accomplish the links between the nodes are presented as a communication medium, carrying out data transfer. This picture will make it easy to follow the client requests and server responses when the discussion is focused at the GN model of the DFS.

A common approach to maintaining high availability and short response time is the use of 'replicated' files. These are files that have copies at multiple nodes across the distributed system. When their modification does not take place frequently, this method is quite efficient. However, attempts to modify a replicated file invoke special synchronization mechanism, which causes heavy communication traffic and affects system performance. Ordinary files do not have multiple copies and are processed in a simpler way, but their average access time is usually higher.

Normally a DFS allows several operations to be carried out

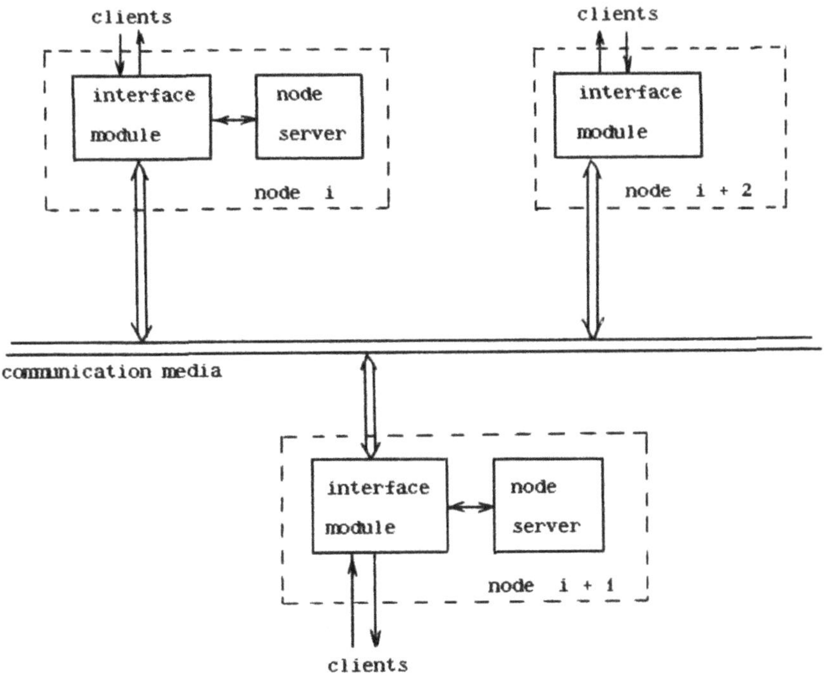

Fig. 5.1

upon files, such as read, write, create, etc. There is a special way in which all file modifications take place: they are either completed successfully, or the system is left unchanged. A specific term is associated with this type of action, reflecting its nature - 'atomic transaction'.

Taking into consideration the above features of a DFS, it has been possible to form a GN model of the distributed system currently designed at the Institute of Microsystems, Sofia. This model is shown in Fig. 5.2. The places in the net represent the states, in which client requests appear, while they are being serviced by the DFS. The token stands for a particular client request at a given time. The transitions of these tokens between the places are determined by the respective predicates.

An important advantage of GNs is their ability to assign certain characteristics to the tokens. They may be further used to analyse the events that have taken place in the DFS, their causes and results.

Some definitions have to be made prior to their use in predicates and token characteristics. Let every file 'f' in the system have a function 'r(f)' defined for it, such that

$r(f) = 1$, if 'f' is an ordinary file, and

$r(f) > 1$, if 'f' is a replicated one.

On this basis, the sets 'Fo' and 'Fr' of ordinary and replicated files in the DFS are defined as follows:

$$Fo = \{ f \,/\, r(f) = 1 \},$$

$$Fr = \{ f \,/\, r(f) > 1 \}.$$

Every token 'α' entering the net is assigned the initial characteristic x_0^{α} = "<client_node, f, o>", where 'client_node' is the node in which the respective request is serviced, 'f' is the file identifier, and 'o' is the operation to be applied on 'f'.

In places l_3 and l_5 the tokens are assigned new characteristic expressed by the respective characteristic functions Φ_3 and Φ_5:

Φ_3 -> "<server_node>",

where 'server_node' is a node in the system where the request is to be serviced.

Φ_6 -> "<trans_id, TIME, ∂, result>",

where 'trans_id' is a transaction identifier assigned by the system; 'TIME' is the current time, '∂' is the period during which the transaction is active, 'result' signifies whether the operation has or has not been successful:

$$\text{result} \in \{ \text{ true, false } \}.$$

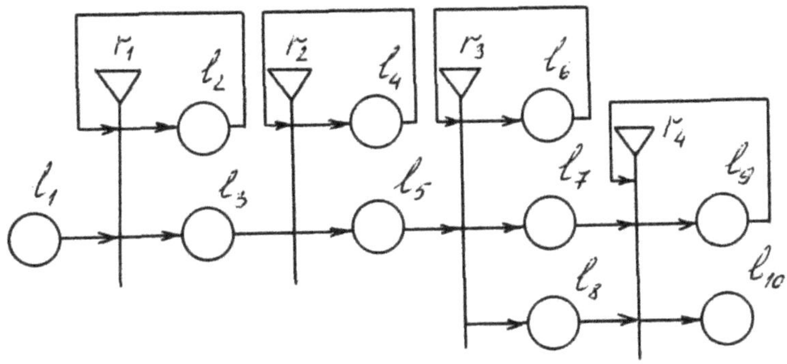

client request | begin transaction | abort transaction | server response

l_2 -replicate $\quad$ l_6 - synchronize $\quad$ l_9 - remore response

l_4 - remote request $\quad$ l_7 - end transaction

Fig. 5.2.

The transition conditions are defined as follows:

$$r_1 = \begin{array}{c|cc} & l_2 & l_3 \\ \hline l_1 & W_1 & W_2 \\ l_2 & W_3 & \text{true} \end{array}$$

where

W_1 = "$f \in Fr$";

W_2 = "$f \in Fo$";

$W_3 = $ "$\bar{\bar{c}}(l_3, X_0^{\alpha}) < r(pr2\ x_0^{\alpha})$"

where $\bar{\bar{c}}(l, X)$ being a function giving the number of tokens in place 'l' with characteristic 'X';

$r_2 =$

	l_4	l_5
l_3	W_4	W_5
l_4	false	true

where

$W_4 = $ "$x_{cu}^{\alpha} = pr_1\ x_0^{\alpha}$";

$W_5 = \neg W_4$;

$r_3 =$

	l_6	l_7	l_8
l_5	W_6	W_7	W_8
l_6	W_9	W_{10}	W_{11}

where

$W_6 = $ "$f \in Fr$";

$W_7 = $ "$f \in Fo$" & "$pr_3\ x_{cu}^{\alpha} \le Tr$" & "$pr_4\ x_{cu}^{\alpha} = true$"

where 'Tr' being the maximum time allowed by the system during which a transaction may be active,

$W_8 = $ "$f \in Fo$" & "$pr_3\ x_{cu}^{\alpha} > Tr$" v "$pr_4\ x_{cu}^{\alpha} = false$";

$W_9 = $ "$\bar{\bar{c}}(l_6, x_0^{\alpha}) < r(pr_2\ x_0^{\alpha})$";

$W_{10} = $ "$\bar{\bar{c}}(l_6, x_0^{\alpha}) = r(pr_2\ X_0^{\alpha})$" & "$pr_3\ x_{cu}^{\alpha} \le Tr$" & "$pr_4\ x_{cu}^{\alpha} = true$";

$W_{11} = $ "$\bar{\bar{c}}(l_6, x_0^{\alpha}) = r(pr_2\ X_0^{\alpha})$" & "$pr_3\ x_{cu}^{\alpha} > Tr$" & "$pr_4\ x_{cu}^{\alpha} = false$";

$$r_4 = \begin{array}{c|cc} & l_9 & l_{10} \\ \hline l_7 & W_{12} & W_{13} \\ l_8 & W_{12} & W_{13} \\ l_9 & false & true \end{array}$$

where

$W_{12} = \neg W_{13}$;

$W_{13} =$ "$x_1^{\alpha} = pr\ x_{10}^{\alpha}$".

The presented GN model of a DFS does not describe all the possible situations in a full-scale system, which is rarely needed. Usually during the time when a DFS is designed, particular assumptions are made and some restrictions are imposed. This is done to prevent the possibility of eventual indefinite states in the system, to avoid deadlocks and other undesired effects. A predictful behaviour is often achieved after a reasonable compromise between the complexity of a model and the clarity of its operation.

In this particular case some potential events are ignored, such as communication failures, invalid client requests or access rights. These situations are easily solvable, but the model would be unreasonably complicated and will not be adequate for its present purpose.

Further research work in the field of DFS design will probably bring forth other problems, and methods for their solution. The author's belief is that a precise GN model of any distributed file system could suggest helpful ideas about its design and development.

§ 5.2: GENERALIZED NET MODELS OF DATA TRANSFER IN INDUSTRIAL LOCAL AREA NETWORKS

Alexander Savov

§ 5.2.1 Introduction

This paper by means of the Generalized Nets (GNs) presents the results of a simulation study carried out by an Industrial Local Area Network (ILAN). The results are very close to the designed specifications. The ILAN is composed of three layers: physical, data link, and application [1]. The data link layer is built on a polling mechanism because of its deterministic behavior. All stations are polled either centrally by a *supervisor or decentrally by a token passing protocol* [2]. There are two types of stations: initiators and responders. Initiators provide the following data transfer services [3]:

(a) SDA - send data with acknowledgement,

(b) SDN - send data with no acknowledgement,

(c) RDR - request data with reply.

The initiators connected to the bus cooperate by using the shared channel. The right to use the channel is named a token. A token is passed from a station to a station in a cyclic way, thus defining a logical ring.

The services SDA and RDR are a couple of frames sent via the medium: the frame - request - passes from the initiator to the responder, the frame - answer - passes from the responder to the initiator.

The service SDN has only one frame request passed from the initiator to the responder.

The stations included in the ILAN are Programmable Devices (PDs) - components of manufacturing systems such as programmable logic controllers, robots, process controllers, personal computers, and others. Their base functions are calculating the control programme and updating inputs and outputs. The control programme works cyclicly. The cycle is called a Treatment Cycle (TC). The PD takes inputs when the TC starts. This is done in order to avoid hazards due to value changes during a cycle. For the same reason data exchange is impossible when the TC runs. Two models of ILAN exist:

(a) model 1 - only initiators are included in a logical ring,

(b) model 2 - all stations are included in a logical ring.

<u>Model 1 behaviour</u>

1. The initiator receives the token and sends the frame - request to the responder.
2. The receipt of this frame causes the responder to wait the TC to end, to compose and transmit the frame- answer.
3. The receipt of this frame causes the initiator to pass the token to another initiator.

<u>Model 2 behaviour</u>

1. The initiator receives the token, sends the frame - request to the responder, and passes the token to the next initiator.
2. The receipt of this frame causes the responder to wait for the TC to end, to compose the frame-answer, and to wait for the token.
3. The responder receives the token and transmits the frame - answer.

§ 5.2.2 Generalized net model

By means of GNs we shall describe a GN-model (see Fig. 5.3) of data transfer in ILAN.

N is the number of stations included in the ILAN.

I is the number of initiators included in the ILAN.

Lmin is the minimum length of the data field that can be trasferred by one service.

Lmax is the maximum length of the field of data that can be trasferred by one service.

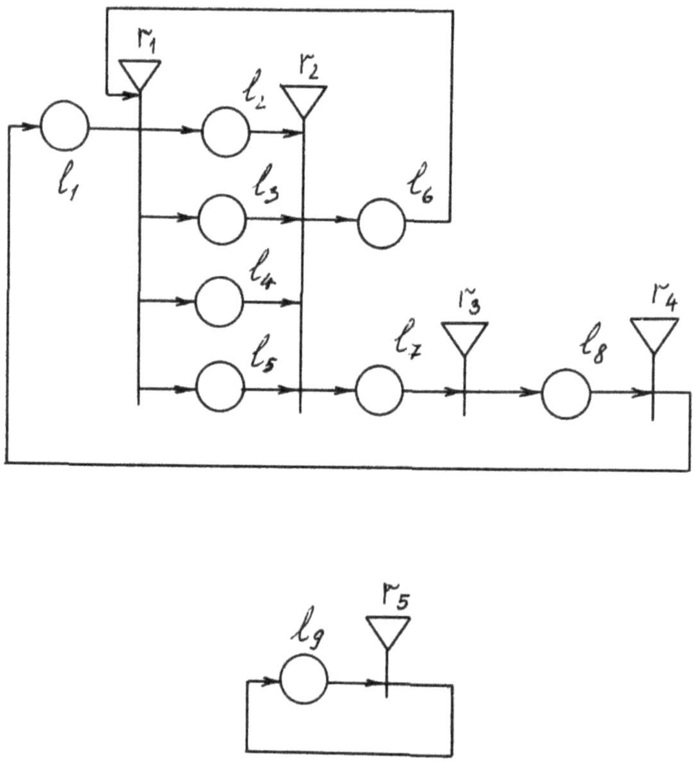

Fig. 5.3

CIi is a TC in the initiator when it composes a frame-request.

CIi is a TC in the initiator when it processes a frame-answer.

CR is a TC in the responder when it processes a frame-request and proposes a frame-answer.

The GN-model consists of two parts. The first models the functions of the ILAN and the second (transition r_5, place l_9 and token α) - the determined protocol on the data link level.

The tokens $\beta_1, \ldots, \beta_N$ in the GN describe the transfer and processing of the data in the media and the station.

The tokens in the first part of the GN have the following initial characteristics: <SA, DA, L, type>, where

(a) SA is the address of the station initiator, the source of the service;

(b) DA is the address of destination station - responder, the receiver of the service;

(c) L is the length of the data field that is transferred during the service;

(d) type is the type of service (SDA, RDR, SDN [2]).

These tokens pass consequently through the places l_8, l_1, l_2, l_6, l_3, l_6, l_4, l_6, l_5, l_7 and l_8.

One service is modelled by passing a token through the net. The places model the following:

l_1 - setting up of the frame-request in the initiator;

l_2 - frame-request's buffer in the initiator;

l_6 - transmission of the frame-request and frame-answer in the media;

l_3 - processing of the frame-request and setup of the frame-answer's buffer in the responder;

l_4 - frame-answer's buffer in the responder;

l_5 - no physical meaning;

l_7 - processing of the frame-answer in the initiator.

The time intervals, when the tokens are in the different places, depict the processing of the data in the stations and the time of transmission in the media. These time intervals are a function of the length of the data-field and depend mostly on

(a) the speed of the microprocessors and the hardware in the data-transfer section (DMA or other),

(b) the data transmiting speed in the media,

(c) the frame organization and the data representation.

The following equations are used for estimation of the duration pf a tokens' stay a certain place:

(a) for l_1 : T1 - the time for setting up the frame-request

T1 = K1+L. t1+R. CI1 (for the services SDA, SDN),

T1 = K1+R. CI1 (for the service RDR),

where

K1 - a constant,

t1 - a time constant depending on the speed of the hardware,

R - a random number in the interval [0, 1],

(b) for l_6 : T6. 2 - time for transmitting the frame-request

T6. 2 = K2+S. t2 (for the service RDR),

T6. 2 = K2+(S+L). t2 (for the services SDA, SDN),

T6. 4 - time for transmitting the frame-answer:

T6. 4 = K2+S. t2 (for the service SDA)

T6. 4 = K2+(S+L). t2 (for the service RDR)

T6. 4 = 0 (for the service SDN),

where

K2 - a constant,

t2 - a time constant, depending on the speed and the method of the media transmission,

S - the number of the service characters in the frame,

(c) for l_3 : T3 - process time of the frame-request and setting up of the frame-answer in the responder

T3 = K3+L. t3+R. CR;

(d) for l_7 : T7 - process time of the frame-answer in the initiator

T7 = K1+L.t1+R.CI2 (for the service RDR)

T7 = K1+R.CI2 (for the services SDA, SDN)

(e) for l_8 : T8 - time for generation of a request,

T8 = R.t4,

where t4 is a time constant.

The token α in the second part of the net has a characteristic "a" which determines the number of the stations that have an access to the media. "a" is changed on every firing of the transition.

(f) for l_9 : T9 - token holding time,

for model 1:

T9 = t5 if a ≠ SA,

T9 = t5+T6.2+T3+T6.4 if a = SA;

for model 2:

T9 = t5 if a ≠ SA and a ≠ DA,

T9 = t5+T6.2 if a = SA,

T9 = t5+T6.4 if a = DA,

The transition conditions in the GN for model 1 are

$$r_1 = \begin{array}{c|cccc} & l_2 & l_3 & l_4 & l_5 \\ \hline l_1 & W_1 & false & false & false \\ l_6 & false & W_2 & W_3 & W_4 \end{array}$$

where

W_1 = "the time parameter of the token is ≤ the current time",

W_2 = "the token has been in place l_2 before going to 16" &

"the time parameter of the token is ≤ the current time",

W_3 = "the token has been in the place l_3 before going to 16" &

"the time parameter of the token is $\leq$ the current time",

W_4 = "the token has been in the place l_4 before going to 16" &

"the time parameter of the token is $\leq$ the current time";

$$r_2 = \begin{array}{c|cc} & l_6 & l_7 \\ \hline l_2 & W_5 & false \\ l_3 & W_1 & false \\ l_4 & true & false \\ l_5 & false & true \end{array}$$

where

W_5 = "a = SA for the token",

$$r_3 = \begin{array}{c|c} & l_8 \\ \hline l_7 & W_1 \end{array}$$

$$r_4 = \begin{array}{c|c} & l_1 \\ \hline l_8 & W_1 \end{array}$$

$$r_5 = \begin{array}{c|c} & l_9 \\ \hline l_9 & W_1 \end{array}$$

The predicates in the transition conditions for model 2 are similar to the above-mentioned ones with the following difference:

$$r_2 = \begin{array}{c|cc} & l_6 & l_7 \\ \hline l_2 & W_5 & false \\ l_3 & W_1 & false \\ l_4 & W_6 & false \\ l_5 & false & true \end{array}$$

where

W_6 = "a = DA for the token".

The characteristic functions give to the tokens at different places characteristics which are related to the time-stay of the tokens at these places.

The models so described are simulated. There are certain results for the mean time of a service, the optimum traffic and the length of the buffers in the different ILANs.

REFERENCES:

[1] ISO - Open System Interconnection (DIS-ISO 7498)

[2] ANSI/IEEE Std. 802.4-1985. Token passing bus access method and physical layer specifications.

[3] IEC - Process data highway - PROWAY - A, B, C.

§ 5.3: GENERALIZED NET - MODELS OF FFT AND TRANSPUTER SYSTEMS

Antonia Dimitrova and Rossen Petrov

Many signals and phenomena exist in nature which could be described and explained by the power analytical apparatus of the transformation named after the notorious French mathematician Jean Batiste Joseph Fourierr. By the Fourier analysis an arbitrary spatial or time function can be decomposed into sinusoidal constituents, each of which possesses its own amplitude, frequency, and phase. The transformation represents a transition from the time area to the frequency area; it is a basis for solving many research problems. In essence, the Fourier transformation performs a spectral analysis of the examined values. A basic goal of the implementation of the Fourier transformation is minimizing the time for its computation. For this purpose a variant for accelerated transformation computation was developed by Cooley and Tukey in 1965, as Fast Fourier Transformation (FFT). By computation of the FFT upon K discrete readings (reports, samples), the acceleration which is obtained by the FFT implementation compared with the standard Fourier transformation is $K/\text{Log}_r K$, where r is the radix of the FFT. The radix of the FFT could be 2, 4, 8, 16, etc. The most widely used variant is the one with radix 2. In this case the number of the processed readings is an exact power of 2. A typical feature of the FFT algorithm is its implementation in m consequent phases, where $m = \text{Log}_2 K$. The Fourier transformation is applied to both real and complex input signals. The output signal is always complex. All the intermediate results are complex values too. These peculiarities (features) set up grave requirements on computational environments in which FFT is implement. That is why the problem for parallel implementation of the FFT with a maximum load of parallel working computational cells is topical and very important. A detailed analysis of the FFT algorithm shows nonlinear dependency of the algorithm acceleration as a function of the number of parallel processors. The aim of the built up model is to set up the number of parallel working processes which will accelerate the FFT algorithm in a maximum way. As a computational environment an INMOS IMS T800 transputer network is chosen. The number of transputers building up the transputer network is 2 to 16. The basic features of these processors, concerning the built model, are as follows:

(a) four serial communication channels for connection with processors of the same type,

(b) independency of the data transfer on the computational process;

(c) built in device for floating point arithmetic with 10 MFLOPS performance;

(d) tact frequency 20 MHz.

Let us have possibility to include p transputers in the computational network. In the common case, it is desirable for p to be a power of 2 due to the fact that the algorithm is implemented by using radix 2 in order that a maximum and uniform load of the computational network is obtained. In this presentation we have a maximum of 16 computational cells which are connected under a rule each by each. Because of the large period for the data transfer, commensurable with the computational time, a large number of computational cells does not correspond to the great speed. This sets up the considered problem. It has been examined that for the considered problem increasing the numbet of cells to over 16 is not effective and this is the reason why such cases not to be considered here.

Fig. 5.4 represents a generalized net. Using it we shall solve the problem about the number of the necessary computational cells to achieve a maximum performance. The data volume is significant when this problem is discussed. That is why it is set up as an initial feature of the token in place l_1. By changing this feature, a new initialization of the network is necessary. In addition to the token in place l_1, another token in place l_8 exists and it has a "maximum time for problem execution" feature. This "maximum time" depends on the will of the operator. For instance, it can be taken from the given task, e.g. some boundary time.

The first token represents a problem in its initial look: the FFT which is to be implemented on K points, over p transputers ($p = 2^i$ for $0 \le i \le 4$) and it is required to determine which variant has the maximum speed. After the transition Z_1 this problem is split into N subproblems, where $N = (p - 1)^2$, as on each of the subproblems a serial number is set in order that they be distinguished from each other. Each subproblem is given a priority in so that the next token does not pass the transition which has as the input both the next and previous tokens.

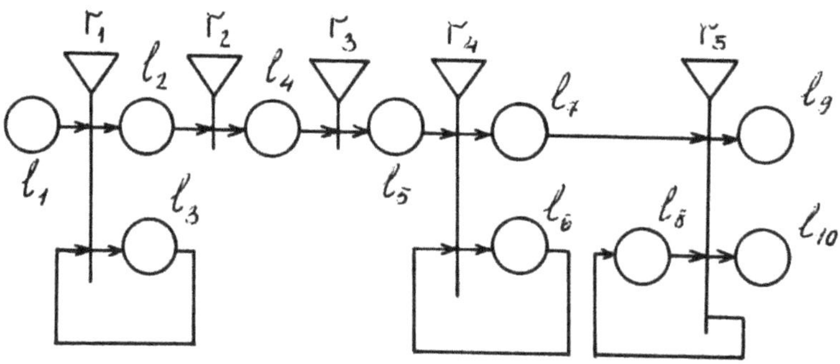

Fig. 5.4

The first token possesses the highest priority and the priority is reduced consequently to the last token. The transition by which such a situation can arise is Z_4.

In the transition Z_5 the features of the tokens in places l_7 and l_8 containing the total subproblem execution time are compared. As a result, in the place l_9 the token which has the least execution time will be obtained and the problem will be solved at this moment.

$$Z_1 = \langle \{l_1, l_3\}, \{l_2, l_3\}, r_1, M_1, \vee(l_1, l_3)\rangle,$$

where

$r_1 =$

	l_2	l_3
l_1	false	W_1
l_2	true	false

where

W_1 = "The number of the firings up to the moment, $n < N$";

$Z_2 = \langle \{l_2\}, \{l_4\}, r_2, M_2, \vee(l_2)\rangle$;

$Z_3 = \langle \{l_4\}, \{l_5\}, r_3, M_3, \vee(l_4)\rangle$;

where the transition condition predicates of r_1 and r_2 are "true";

$Z_4 = \langle \{l_5, l_6\}, \{l_6, l_7\}, r_4, M_4, \vee(l_5, l_6)\rangle$

where

$r_4 =$

	l_6	l_7
l_5	W_2	W_3
l_6	false	true

where

W_2 = the number of transputers is > 1",

$W_3 = \neg W_2$;

$Z_5 = \langle \{l_7, l_8\}, \{l_8, l_9, l_{10}\}, r_5, M_5, \wedge(l_7, l_8)\rangle$

where

$$r_5 = \begin{array}{c|ccc} & l_8 & l_9 & l_{10} \\ \hline l_7 & W' \vee W'' & W' \vee W''' & \neg W' \,\&\, \neg W'' \\ l_8 & \text{false} & W''' \,\&\, \neg W' & W' \end{array}$$

where

W' – "The last characteristic of the token in place l_7 is less than the last characteristic of the token in place l_8",

W'' = "The transition firings are equal to 0",

W''' = "The transition firing number is N".

All elements of M_i $(1 \leq i \leq 5)$ are 1. For every token α the following (consecutively) is valid:

$$\pi_K(\alpha) = \{\mathrm{Log}_2\, p+1,\ \mathrm{Log}_2\, p,\ \ldots,\ 1\}.$$

Tokens enter with the initial characteristic "Data volume, k" and they receive the following ones:

Φ_3 -> "current number of transition firings";

Φ_4 -> "$k.T_t$", where T_t is the one data item transfer time",

Φ_5 -> "$k.T_c.((\mathrm{Log}_2 k) - N + 1)/(2.(N-1)^2)$", where T_c is the basic operation computation time;

Φ_6 -> "$k.(T_t + T_c)/(2*(N-1)^2)$;

Φ_7 -> "$k.T_t$";

Φ_9 -> "$(n-1)^2$".

§ 5.4: IMPLEMENTATION OF GENERALIZED NETS FOR SIMULATION OF NODE-TO-NODE COMMUNICATION IN A REGULAR STRUCTURE

Stefan Stefanov

The case of node-to-node communication in a massively linked network of nodes is discussed.

A structure which we shall call a massively linked regular graph (MLRG) is shown in Fig. 5.5.

Each node must have 8 bidirectional links. In case the hardware has only 6 or 4 bidirectional links each node may be replaced by a cluster like the ones shown in Fig. 5.6. In that case we call the structure a massively linked regular cluster graph (MLRCG).

Bearing in mind that

a) the nodes are very close to one another in space, and,

b) the main difficulties in communication are the conflicts between the messages trying to pass through the same link at the same time,

the distance (or the message price) between two nodes N_i and N_j may be defined as the number of links which the message passes through in order to reach N_j starting from N_i.

The distance (message price) between N_i and N_j is

$$P(N_i, N_j) = \max(|v(N_i) - v(N_j)|, |h(N_i) - h(N_j)|),$$

where $v(N_i)$ is the relative vertical position of N_i in the graph and $h(N_i)$ is the relative horizontal position. Note that this result is optimal (even better than could be expected).

Let D = {E, W, N, S, NE, NW, SE, SW} be the set of the possible "world" directions. Let us substitute

$$v(N_i) - v(N_j) = v$$

$$h(N_i) - h(N_j) = h.$$

Then $P(v, h) = \max(|v|, |h|)$.

The primary direction calculation mechanism (PDCM) is given in the table below, where

$$sgn(x) = \begin{cases} "-", & \text{if } x < 0 \\ "0", & \text{if } x = 0 \\ "+", & \text{if } x > 0 \end{cases},$$

$$d(v, h) \in D,$$

"NONE" means that no direction is calculable.

sgn(v)	sgn(h)	d(v, h)
0	0	None
0	-	W
0	+	E
-	0	N
-	-	NW
-	+	NE
+	0	S
+	-	SW
+	+	SE

The secondary direction calculation mechanism type of A (SDCMA) is defined recursively:

applicable if $h \neq 0$ and $v \neq 0$ and $v \neq h$

$$d'A(v, h) = \begin{cases} \text{if } (|v| > |h|) \text{ then } d(v, 0) \\ \text{if } (|v| < |h|) \text{ then } d(0, h) \end{cases}.$$

The secondary direction calculation mechanism type of B (SDCMB) is given by the table below which is based on the PDCM.

The "nearest possible direction" to d(v, h), where $d'B(v, h) \in D$, is defined as follows:

d(v, h)	d'B(v, h)
NE	N or E
E	NE or SE
SE	E or S
S	SE or SW
SW	S or W
W	SW or NW
NW	W or N
N	NW or NE

The tertiary direction calculation mechanism (TDCM) determines

$$d''(v, h) \in D - \{d(v, h)\},$$

i.e., d"(v, h) is any member of D different from d(v, h).

In the case of a MLRCG each node is a cluster and the in-cluster directions are calculated similarly (using tables).

Note that if the SDCMA is chosen, instead of the PDCM, the message's price P does not increase, i.e. an optimal routing is achieved if the PDCM or the SDCMA, is used.

If the SDCMB is chosen instead of the PDCM then P is increased by 1.

If the TDCM is chosen instead of the PDCM then P is increased by more than 1.

Tertiary directions are chosen in emergency cases (hardware or software malfunctions).

A problem that is solved by the simulation is the following.

Let C be the set of the vertices of the graph, $Cs \subset C$ be the set of message senders, $Cd \subset C$ be the set of message destinations,

$$n1 = |Cs|,$$

$$n2 = |Cd|.$$

The time W = W(C, n1, n2) during which all messages wait to reach their destinations is a topic of importance.

The generalized net (GN) representing the case is shown in Fig. 5.7.

The model consists of 7 transitions and 11 places. It can be further minimized, but this is not done to preserve clarity and the closeness to physics of the process.

Tokens from the initial place L_i pass through the trivial transition T_1 where they receive their initial location coordinates $cc \in Cs$, destination coordinates $cd \in Cd$, with priority equal to the time when they enter the transition, Wc and Wg being zeroes and all the other elements of the characteristic vector blanks.

Transition T_2 stays inactive until all ni tokens gather in L_1. Passing through the trivial T_2 a token receives its next partial destination $cn \in C$, calculated using the PDCM. Then it continues through T_4 and goes either to $L_{7,i}$ ($1 \le i \le K$, and K is the number of nodes in the graph) if it is empty, or to L_8 if otherwise. From $L_{7,i}$, tokens go to L_9 and from there either to L_o if cc = cd, or to L_2 to continue moving in the net.

A token from L_8 goes either to L_5 to retry to reach its partial destination, or, if it has tried for more than X times, to L_6. Passing through T_3 a new partial destination $cn \in C$ is calculated using the SDCM. The simulation stops when all tokens gather in L_o. X is the maximum retry count.

The time-parameters of the net are: $T = 0$, $t^o = 1$, $t^* = \infty$.

The transition conditions are

$$r_1 = \begin{array}{c|c} & L_1 \\ \hline L_I & W_1 \end{array}$$

where

W_1 = "TIME > n1";

$$r_2 = \begin{array}{c|c} & L_3 \\ \hline L_1 & true \\ L_2 & true \end{array}$$

$$r_3 = \begin{array}{c|c} & L_4 \\ \hline L_6 & true \end{array}$$

$$r_4 = \begin{array}{c|ccccc} & L_{7,1} & L_{7,2} & \dots & L_{7,N} & L_8 \\ \hline L_3 & W_2 & W_2 & \dots & W_2 & W_3 \\ L_4 & W_2 & W_2 & \dots & W_2 & W_3 \\ L_5 & W_2 & W_2 & \dots & W_2 & W_3 \end{array}$$

where

W_2 = "$\bar{c}(L_{7,i}$, TIME) = 0" & "Cn = $c(l_{7,i})$",

where "i" is the number of the current place;

$W_3 = \neg W_2$;

$$r_5 = \begin{array}{c|cc} & L_5 & L_6 \\ \hline L_8 & W_4 & W_5 \end{array}$$

where

W_4 = "W_c < X",

$W_5 = \neg W_4$;

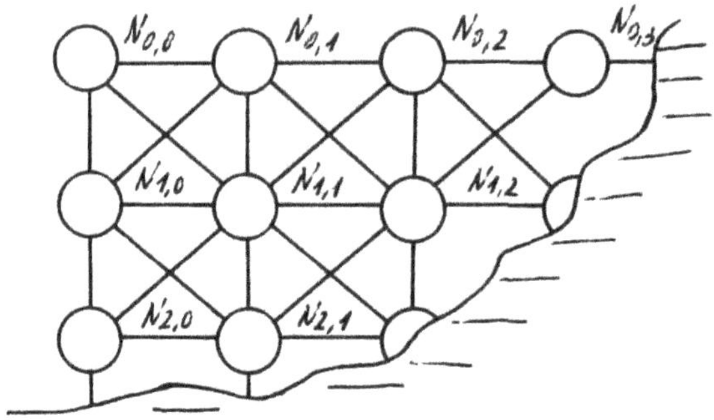

Fig. 5.5.

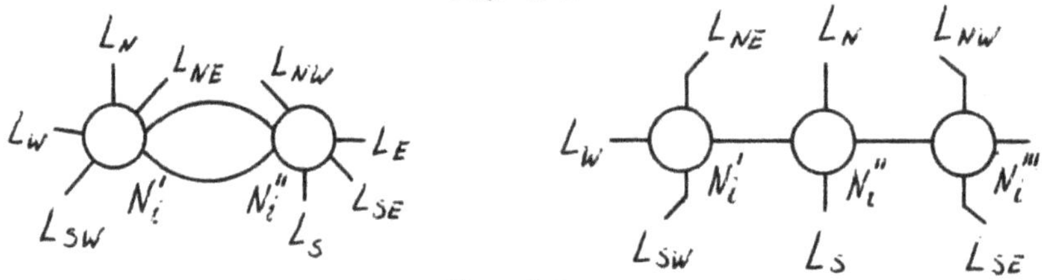

Fig. 5.6.

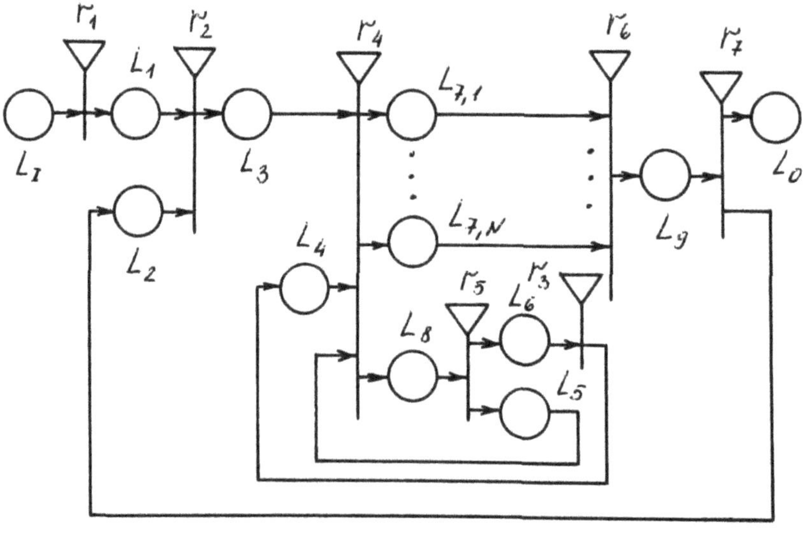

Fig. 5.7.

$$r_6 = \begin{array}{c|c} & L_{7,1} \\ \hline L_{7,1} & true \\ L_{7,2} & true \\ \vdots & \\ L_{7,N} & true \end{array}$$

$$r_7 = \begin{array}{c|cc} & L_0 & L_2 \\ \hline L_9 & W_6 & W_7 \end{array}$$

where

W_6 = "$C_c = C_d$";

$W_7 = \neg W_6$.

Every token has only one (its current) characteristic "<cc, cd, cn, Wc, Wg>", where

(a) cc are the current coordinates;

(b) cd are the destination coordinates;

(c) cn are the coordinates of the next partial destination;

(d) Wg is the global wait time (retry count);

(e) Wc is a scratch (current retry count).

The characteristic functions related to the places are

Φ_1 -> "<cc, cd, cn, Wc, Wg>",

where

(a) cc is the initial location (the message source);

(b) cd is the destination;

(c) cn is undefined;

(d) Wc is 0;

(e) Dg is 0;

Φ_3 -> "<cc, cd, cn, Wc, Wg>";

where "cn" is the next partial destination calculated using the PDCM, all the rest remaining unchanged;

Φ_4 = "<cc, cd, cn, Wc, Wg>";

where "cn" is the next partial destination calculated using the SDCM (A or B), all the rest remaining unchanged;

Φ_5 = "<cc, cd, cn, Wc, Wg>";

where

(a) Wc = Wc + 1;

(b) Wg = Wg + 1;

all the rest remaining unchanged;

Φ_6 = "<cc, cd, cn, Wc, Wg>";

where

(a) Wc = 0;

(b) Wg = Wg + 1;

all the rest remaining unchanged;

Φ_9 = "<cc, cd, cn, Wc, Wg>";

where cc = cd, all the rest remaining unchanged;

Φ_2 -> "*";

$\Phi_{7,i}$ -> "*";

Φ_8 -> "*";

Φ_0 -> "<Wg, TIME - n1>",

(this is the final token's characteristic; TIME is the current model time).

All the places and arcs have infinite capacities. All the transitions and places have equal priorities. All transition types are "v".

§ 5.5: GENERALIZED NET MODELLING OF DISK SUBSYSTEMS

Bistra B. Nikolova

Disk Drives (DDs) and the subsystems built on their bases, are one of the most important components of the computer. They are used for expansion of the RAM of the computer. The productivity of the computer depends on them.

An example of a structure scheme of a disk subsystem (DS) is shown in Fig. 5.8. It consists of an input/output path (i/o path) and M numbers of DDs. The whole software and hardware, necessary for the data exchange between the RAM and DDs is represented as a separate functional unit, placed between the RAM and DDs [1]. This unit is marked as an i/o path. If a certain part of the hardware, which takes part in the i/o path of the concrete subsystem, is occupied by a DD, the other DD cannot exchange data. Data exchange is realized by direct memory access.

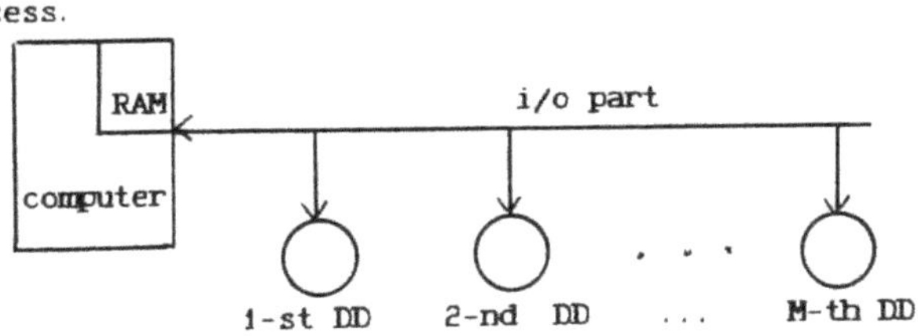

Fig. 5.8.

There are some kinds of architectures in which in order to increase the parallelity in the functioning of the DS, new i/o paths are added. They are called "subsystems with a dynamic distribution of the control between a certain number of i/o paths". They consist of M numbers of DDs and N numbers of i/o paths, $2 \le N \le M$ [4]. Every DD is connected to every i/o path. The control of the i/o operations with DDs and their parts can be done with the help of as i/o path. The requests to an i/o path form a queue [2, 5].

Usually, the i/o path is used for exchange, not only of data, but also of control information. If a separate path for commands is added, two buses are formed for data and commands. The opportunity for positioning at every moment of the DD is created, nevertheless the data bus and the i/o path are busy with the data exchange from/to a DD.

The result of an i/o operation is given in Fig. 5.9, where

A - the request goes into the DS (in the queue to the DDs)

B - the i/o operation is started

C - the positioning of the DDs is completed and the request goes into the queue of the i/o path

D - the data exchange begins

E - the request service ends and the request leaves the DS

T_{DQ} - the time during which the request is in the queue to the DDs

T_{DS} - position time

T_{CQ} - the time, during which the request is in the queue to i/o path

T_{CS} - data exchange time

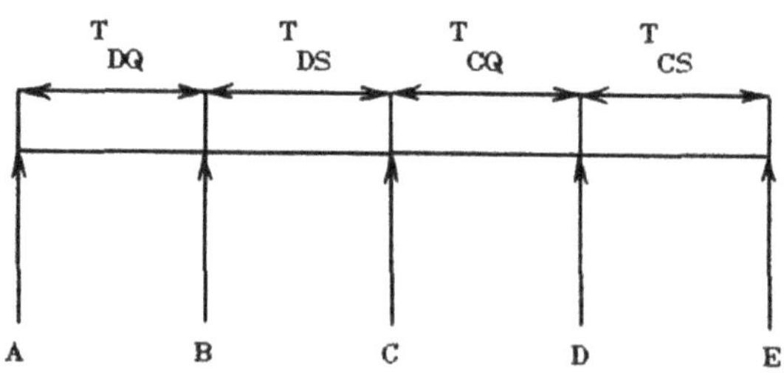

Fig. 5.9.

On the other hand, an i/o operation in the DS consists of positioning, searching and exchanging data.

The purpose of the article is to make a model of some DS architecture in terms of Generalized Nets (GNs).

Some specifications are made:

1) all DDs are equal and the input requests are equally distributed between them [2, 5];
2) the fulfilment of the i/o operations includes the steps from Fig. 5.9 [2, 5];
3) the queues to the DDs and the i/o path are realized by means of the FIFO [2, 3];
4) every D has its own queue of requests, while the queue to the i/o path is common;
5) DD positions the head over the wanted cylinder by itself;
6) the i/o operation starts from a free DD and i/o path; otherwise the request moves either to the queue of the DD or to the queue of the i/o path, according to the architecture of the DS;
7) the i/o path takes part in the search for the block on the track and in the exchange of information between the DDs and the i/o path [3].

The requests in the DS are the tokens in the given model. Each token α goes in the net with an initial characteristic: "<NDS, NCYLN, NB, LDBL, IOO>" (= x_0^α), where NDS, NCYLN, NB, LDBL are natural numbers and

NDS is the number of the DD, $1 \le$ NDS $\le$ MaxNDS, where MaxNDS is the maximum number of the DD,

NCYLN is the number of the cylinder, $0 \le$ NCYLN $\le$ MaxNCYLN, where MaxNCYLN is the maximum number of the cylinder,

NB is the number of the block on the track, $1 \le$ NB $\le$ MaxNB, where MaxNB is the maximum number of blocks on the track,

LDBL is the length of the exchange data in the blocks, $1 \le$ LDBL $\le$ MaxLDBL, where MaxLDBL is the maximum length of the exchange data in the blocks, equal to the information capacity of the DD in the blocks,

IOO is the code of the i/o operation, IOO = <READ, WRITE>, where READ is the code of the read-operation, WRITE is the code of the write-operation.

A request service has been modelled as a transfer of a token through the GN described in Fig. 5.10.

The places in the net used for modelling of different architectures are denoted as follows:

b_0 - models the preparation of requests,

b'_0 - distributes the requests to the corresponding one of the m-th DDs,

b_1 - a queue to the DD,

b_2 - the DD is free,

b'_2 - the DD is not free,

b_3 - the i/o path is free,

b'_3 - the i/o path is not free,

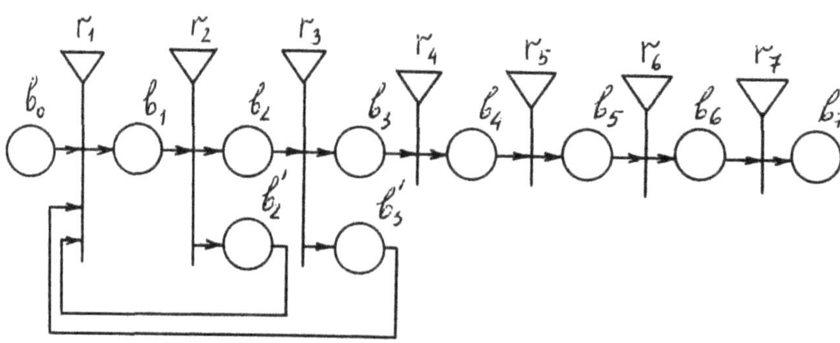

Fig. 5.10.

b''_3 - the independent path of commands is free,

b'''_3 - at least one of the N numbers of the i/o path is free,

b''''_3 - none of the N numbers of the i/o path is free,

b_4 - start of an i/o operation,

b_5 - independent positioning of the DD over the desired cylinder,

b_6 - a queue of the i/o path,

b_7 - a free i/o path and search for a block,

b'_7 - the i/o path is busy and search is not possible,

b''_7 - all the N numbers of the i/o path are busy,

b_8 - exchange between DDs and RAM.

First, we shall discuss a DS which consists of a DD and an i/o path, as is shown in Fig. 5.10. Each token α with the above initial characteristic enters the net. The request goes in the queue of the DDs. This queue is processed by means of a FIFO-discipline.

The GN modelling this architecture is characterized by the following components and parameters. It contains 6 transitions with conditions:

$$r_0 = \begin{array}{c|c} & b_1 \\ \hline b_0 & true \\ b_2 & true \\ b_3 & true \end{array}$$

$$r_1 = \begin{array}{c|cc} & b_2 & b'_2 \\ \hline b_1 & W_1 & W_2 \end{array}$$

where

$W_1 = $ "$\bar{c}(b_2, TIME) = 0$";

$W_2 = \neg W_1$;

(the function $\bar{c}(l, TIME)$ determines the number of tokens in place l at the time-moment TIME);

$$r_2 = \begin{array}{c|cc} & b_3 & b'_3 \\ \hline b_2 & W_3 & W_4 \end{array}$$

where

$W_3 = $ "$\bar{c}(b_3, TIME) = 0$";

$W_4 = \neg W_3$;

$$r_3 = \begin{array}{c|c} & b_4 \\ \hline b_3 & true \end{array}$$

$$r_4 = \begin{array}{c|c} & b_5 \\ \hline b_4 & true \end{array}$$

$$r_5 = \begin{array}{c|c} & b_7 \\ \hline b_5 & true \end{array}$$

$$r_6 = \begin{array}{c|c} & b_8 \\ \hline b_7 & true \end{array} .$$

The characteristic function is defined as follows:

$$\Phi_2 \rightarrow \begin{cases} \text{"}\langle TIME, TIME - \theta_K(\alpha)\rangle\text{", if } \alpha \text{ goes into } b_2 \text{ for the first time} \\ \text{"}\langle TIME, TIME - pr_1 x^{\alpha}_{cu}\rangle\text{", otherwise} \end{cases}$$

where x^{α}_{cu} is the current characteristic of token α;

Φ_3 -> "<TIME, TIME - $pr_1 x^{\alpha}_{cu}$>";

Φ_4 -> "$h_1(pr_2 x^{\alpha}_{cu}, pr_2 x^{\alpha}_{cu-1})$", where the function h_1 determines the time for waiting in place b_4;

Φ_5 -> "$h_2(pr_2 x^{\alpha}_{0}, pr_4 x^{\alpha}_{0})$", where the function h_2 determines the time for positioning;

Φ_7 -> "$h_3(pr_2 x^{\alpha}_{0}, pr_4 x^{\alpha}_{0})$", where the function h_3 determines the time for searching for the block;

Φ_8 -> "$h_4(pr_3 x^{\alpha}_{0}, pr_4 x^{\alpha}_{0})$", where the function h_4 determines the time for exchange between the DD and the RAM.

Second, we shall discuss a DS which consists of an independent command path, an i/o path, and M DDs, as it is shown in Fig. 5.11. The independent command path controls the search and exchange of information. Each token α with the above initial characteristic enters the net. We accept that in practice the independent command path is always free. The queue to the DDs and the i/o path, as above, are processed by means of a FIFO-discipline.

The GN modelling this architecture is characterized by components, which are similar to the above ones.

The transition condition r_0 is

$$r_0 = \begin{array}{c|cccc} & b_{0,1} & b_{0,2} & \dots & b_{0,M} \\ \hline b_0 & W'_1 & W'_1 & \dots & W'_1 \end{array}$$

where

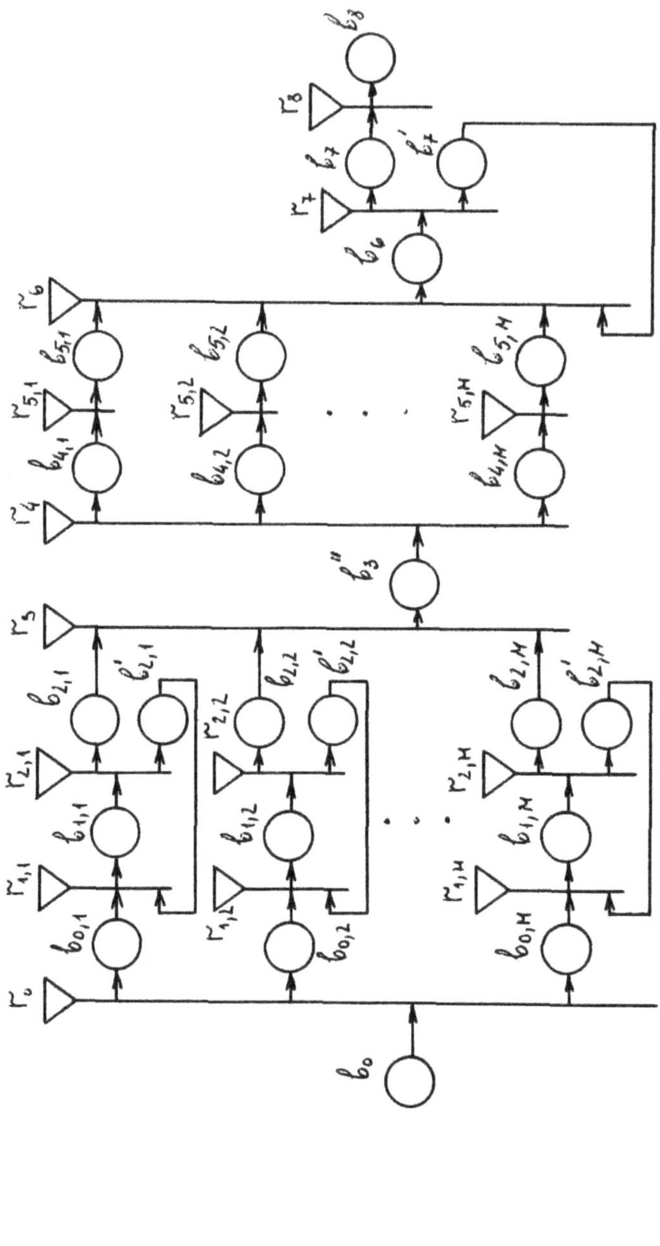

W'_1 = "the number of place is equal to $pr_1 x_0^{\alpha}$".

The transition conditions $r_{1,i}$ are

$$r_{1,i} = \begin{array}{c|c} & b_{1,i} \\ \hline b_{0,i} & true \\ b_{2,i} & true \end{array}$$

for $1 \le i \le M$.

The transition conditions $r_{2,1}$, $r_{2,2}$, ..., $r_{2,M}$ are similar to r_1 with exactly the same notation.

The transition conditions r_3 and r_4 have the forms:

$$r_3 = \begin{array}{c|c} & b''_3 \\ \hline b_{2,1} & true \\ b_{2,2} & true \\ \vdots & \\ b_{2,M} & true \end{array}$$

$$r_4 = \begin{array}{c|cccc} & b_{4,1} & b_{4,2} & \dots & b_{4,M} \\ \hline b''_3 & W'_1 & W'_1 & \dots & W'_1 \end{array},$$

Noting that the capacity of place b''_3 is 1.

The transition conditions $r_{5,1}$, $r_{5,2}$, ..., $r_{5,M}$ are similar to r_5 with exactly the same notation.

The transition conditions r_6 and r_7 have the forms:

$r_6 =$

	b_6
$b_{5,1}$	true
$b_{5,2}$	true
$\vdots$	
$b_{5,M}$	true
b'_7	true

$r_7 =$

	b_7	b'_7
b_6	W_5	W_6

where

W_5 = "$c(\bar{b}_7$, TIME) = 0";

$W_6 = \neg W_5$;

The transition condition r_8 is the same as the one in the first GN. The characteristic functions in both GNs coincide exactly in notation.

Finally, we shall discuss a DS which consists of DDs in number M and i/o paths in number N, as it is shown on Fig. 5.12. The discipline to the DD and the i/o path, as above, is FIFO.

The differences between the second and the third net are in the forms of transitions conditions r_0, r_3 and r_7. They have the forms:

$r_0 =$

	$b_{0,1}$	$b_{0,2}$	...	$b_{0,M}$
b_0	W'_1	W'_1	...	W'_1
b''''_3	W'_1	W'_1	...	W'_1

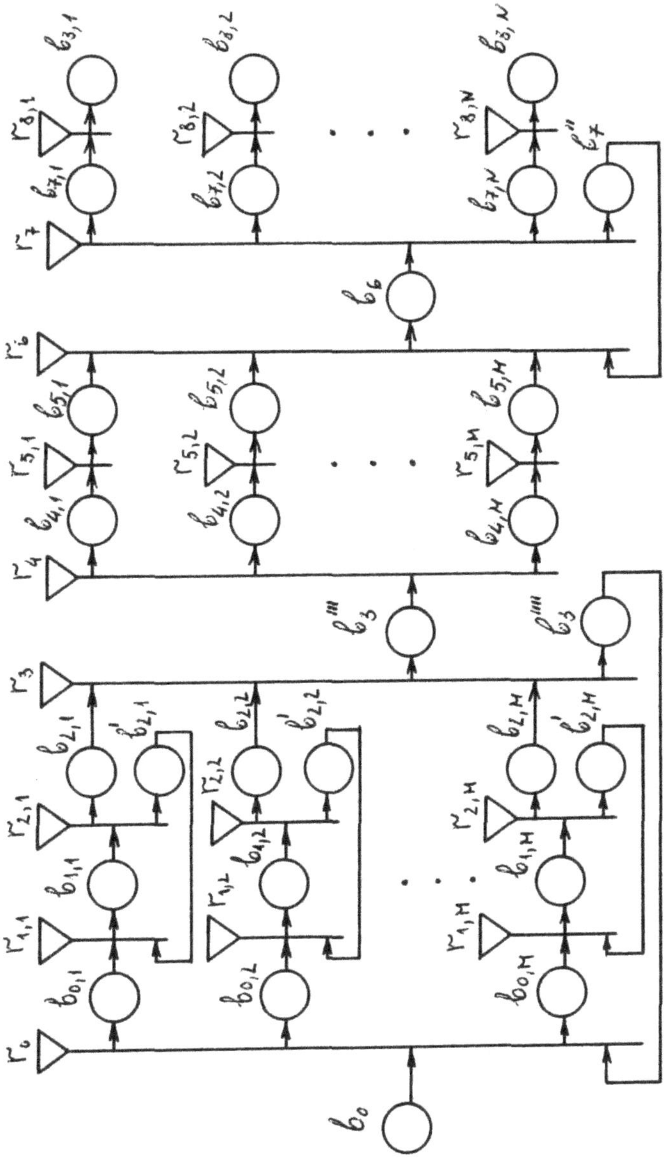

$$r_3 = \begin{array}{c|cc} & b'''_3 & b''''_3 \\ \hline b_{2,1} & W_7 & W_8 \\ b_{2,2} & W_7 & W_8 \\ \vdots & \vdots & \vdots \\ b_{2,M} & W_7 & W_8 \end{array}$$

where

W_7 = "$\bar{c}(b'''_3$, TIME) = 0";

$W_8 = \neg W_7$;

$$r_7 = \begin{array}{c|ccccc} & b_{7,1} & b_{7,2} & \ldots & b_{7,N} & b''_7 \\ \hline b_6 & W_9 & W_9 & \ldots & W_9 & W_{10} \end{array}$$

where

W_9 = "This i/o path is free";

W_{10} = "There is no free i/o path".

The transition conditions $r_{8,1}, r_{8,2}, \ldots, r_{8,N}$ are similar to r_8 with exactly the same notation.

The function Φ is similar to the one from the second GN.

The theory of the GNs makes it possible to create a common formal description of DS.

The concept GN allows all phases of an i/o operation and the dependency between them when i/o requests in the DS are processed to be shown. It also allows all specialities of the separate architectures to be shown.

REFERENCES:

[1] J. Major, Processor, i/o path and DASD configuration capacity, IBM System Journal, Vol. 20, No. 1981. 63-85.

[2] I. Donev, A research over the productivity of disk subsystems,

[3] D. Brown, R. Ei@sen and C. Thorn, Channel and direct access device architecture, IBM System Journal, Vol. 1 No. 3, 1972. 186-199.

[4] R. Wilson, Designers rescue superminicomputers from i/o bottle-neck, Computer Design, Vol. 26, No. 18, 1987, 61-71.

[5] I. Donev, A research over the productivity of disk subsystems. Proc. of Symp. "Electronic and computing technics and technology", May 1987, Sofia, (in Bulgarian).

[6] I. Donev, L. Michov and N. Sinyagina, A research over the organization of highly-productive disk subsystems with stratification of information. Automation Control, Computing Technic and Automated Systems, Sofia (in Bulgarian).

§ 5.6: SOFTWARE TOOLS FOR GENERALIZED NETS

Rumen Christov and Stojan Mihov

§ 5.6.1 Introduction

During the last ten years concurrent processes researches have been developed. Too many big and complex systems do not allow an analytic study. Petri Nets (PNs) and their extensions and modifications are a powerful theoretical tool for the study of the properties of complex concurrent systems. Too many software tools for implementation of the PNs and some of their modifications are produced to make the work of scientists easier. These tools contribute to further development and extension of the practical applications of PNs and their modifications [1].

Each extention of the PNs is designed to solve problems in a determined area. Nowadays a trend to generalization and comparison of mathematical models of processes from different areas is observed. But there is no compatibility between the PNs' extentions. It makes the comparison of the processes' models very difficult.

Generalized Nets (GNs) were defined in 1982. They are a powerful mathematical tool for describing and studing the properties of complex concurrent systems. These nets are a continuation of the trend to make tools for modelling of parallel processes.

GNs are an extension of the PNs. It is proved, that all extensions and modifications of the PNs are particular cases of GNs. This fact proves that each process which may be described and studied by PNs or with their extentions may be described and studied by GNs. This means that GNs are a universal language for describing and studying the processes from different scientific and practical areas.

This fact proves the neccesity of software tools for the implementation of GNs. They are developed by the group, working over GNs and the tools are named STGNs. The STGNs include:

(a) presentation of GN-based hierarchical models of processes;

(b) computer tool support, with which it is possible to create and simulate hierarchical GN;

(c) tools for analysing results of a GN-based model simulation.

§ 5.6.2 GNs - a short description

GNs are a complex mathematical object. Many of their components are presented in the other PNs extentions. However, the GNs are not an aggregate of the components of other types of PNs. All components of GNs have a specific content.

GNs have a static and a dynamic structure, temporal components and a global memory.

Static structure

The static structure includes the net transitions, net places and arcs which connect the places with the transitions.

The information necessary for each transition description is:

(a) a transition identifier, which is a unique name in the net for them;

(b) a transition name (only for user applications);

(c) priority towards other transitions in the GN;

(d) temporal components of the transition: time of the next activation and duration of this active status;

(e) type of the transition;

(f) indexed matrix expressing the transition conditions;

(g) a list of the input places for the transitions;

(h) a list of the output places for the transitions.

The transition type, condition matrix and the functions which assign temporal transition components are included in the dynamic net structure.

The transition type contains the *necessary conditions* for the transition activation. In the GNs' theory the transition type has a specific structure, but in software tools it may be an arbitrary logical expression, which may depend on arbitrary events in the net at the same or a certain previous step of the model simulation.

The information necessary for each place description is

(a) place identifier, which is a unique name in the net for them;

(b) place name (only for user applications);

(c) priority towards other places in the GN;

(d) place capacity - maximum number of tokens which may be in the place in one step from the net simulation;

(e) tokens which are in the place before the GN-model simulation starts.

A place may be

(a) a GN input, if it is a member only of the input places' list for any net transition;

(b) a GN output, if it is a member only of the output places' list for any net transition;

(c) internal to the net, if it is a member of the input places' list for any net transition and is a member of the output places' list for another (eventually the same) net transition.

From each input place for each transition in the GN, arcs to each of its outputs are defined. A capacity is assigned to each arc. For each arc a predicate from the transition condition matrix corresponds.

Dynamic structure

Components of the dynamic structure are tokens, types of transitions, conditions of the transitions and temporal functions of the transitions.

The information necessary for each token description is

(a) a token identifier, which is a unique name in the net for them;

(b) a token name (only for user applications);

(c) priority towards other tokens in the GN;

(d) the time when the token enters the GN;

(e) a list of the net input places, through which a token may enter the GN.

At the moment when the net starts to simulate, the tokens may be in any place in the net or may be waiting to enter the net. In the beginning, the tokens may have a list of any initial characteristics. The token's way through the net depends on these characteristics. When the tokens are going in the net, to the lists of characteristics new characteristics are appended. The characteristics are the greatest difference between GNs and other kinds of PNs. The tokens have an individuality during the whole time of net simulation. This fact means that many of the properties of the concurrent processes may be described by characteristics. It simplifies the static net structure and helps the user to understand the model of the process. The GN-based models are more compact than models based on other kinds of PNs. It allows more detailed models of the studied processes to be made and to get more complex results.

Token characteristics are useful but their software implementation is very difficult. They may not be formalized in their full theoretical definition. Software implementation is the implementation of a class of GNs. However, this class is enough to describe and to study real and theoretical processes and to use these software tools for practical applications.

The characteristics have two aspects. First, they are components of the net PSI-function. It is a definition of the way for calculating token characteristic values. Second, they are the list of values, which a certain token gains when it comes into a certain place. The following types of characteristic values are supported:

- a whole number;

- a real number;
- logical values TRUE and FALSE;
- a string;
- a list, containing an arbitrary combination of values of the above types;
- a matrix of whole or real numbers.

Conditions of transitions

The conditions of net transitions are an indexed matrix of predicates (see App. 2). The matrix elements are indexed by transition input and output place identifiers. Each predicate of these matrices corresponds to one of the transition arcs. The values of the predicates are calculated at each step of the model simulation for each token in the corresponding input place. Predicates may be arbitrary expressions which return the values TRUE or FALSE. They may include relations between all net attributes. A very important limitation is that the predicates cannot depend on further events in the net.

Type of transition

The type of transition is a logical expression. It assigns the necessary set of tokens at the transition input places for regular work of the GN. If the type of transition returns TRUE, then the transition has been activated.

Global memory

The global memory includes:

(a) initial token characteristic values;

(b) characteristic functions, which act on each token at the places corresponding to them;

(c) the number of the characteristic values which are kept when the token is in the net.

When the token is at a net's output place, it owns a list of characteristic values. The list corresponds to the token path in the net. It allows the process to be studied by the history of its components represented by tokens in the GN model.

Hierarchical models based on GNs

GNs allow one to create hierarchical models of the studied processes. It is useful for their complex study. A GN may contain a set of subnets. In the primary GN these subnets are represented by places or transitions. If the subnet has only one input and one output, it is represented by a place. If the subnet has several inputs and several outputs, it is represented by a transition. If a token comes into a certain subnet input, then the subnet simulation starts. The simulation of the primary net and its subnets is synchronized by time towards an absolute time scale. The subnets can use information for the events in the primary net and other subnets during the time of their simulation. As a result the tokens gain the result of the subnet action as a characteristic.

As a further development it is provided that the subnets may be other PNs extentions, as Predicative/Transition Nets or Coloured Petri Nets for example or Neural Networks. This fact means:

- models, based on different PNs' extentions may be compared by their actions and results during the time of their simulation;

- models, already built, may be included as parts of a GN based model;

- if a Neural Networks based model is a subnet of the GNs based model, then each Neural Network may be taught and it may define the action of the primary net.

§ 5. 6. 3. Tools for GN representation

The software tools for implementation of GNs consist of:

- a language for description of GNs based models;

- a graphic editor which allows to construct, edit and update GNs based models;

- a tool for simulation of GNs based models;

- a tools for analysis and representation of the results of the simulation of the model as histogramms, piecharts, etc.

Language for GN description

The models based on GNs are represented by language for a GN description (LGND). The LGND is designed and implemented specially for the description of the GNs. It is a high-level programming language. Its reserved words are notions from the GNs' theory. Besides tools, describing the GN structure the LGND contains tools for the assignment and evaluation of predicates and standard functions for all attributes of the net: current time, time, when a selected transition is active, number of tokens in the selected place and many others.

Construction and editing of GNs

The editor in the STGNs allows the user to construct and edit complex hierarchical GNs. The user works with a GN on a computer, supporting high resolution graphic screen and a mouse. The graphic representation of the GN uses theoretically defined symbols for its components. Each editing operation is immediately reflected on the screen of the computer. The user can increase or decrease some parts of the GN image, so that he can have a global view over the model, created by him.

The user has an access to each net element. He may change its attributes by a menu system. He may move or delete a selected element or may observe the subnet, which this element represents, if the element is a place or a transition.

The STGNs recognizes the structure of GNs. It makes a syntax (by the GN theory) analysis to the GN, created by the user and reports him about errors.

GN simulator

The STGNs includes a GN simulator. This is a tool for simulation of the GN, described by the LGND. The simulation corresponds to the common algorithm. Some variances are allowed. Certain tokens can split and join in certain net places. If the splitting is allowed, when a token may go from one input to several output places, it splits and in the net as many new tokens are born with the same characteristics as their parent as the number of the output places is. When all the children of one parent (tokens from one split level) are in a certain place, they can join to become one token. The new token characteristics will not be a list. They are a net with nodes characteristic of different tokens at the same time moments. In a certain practical application, a communication of different tokens is allowed.

The user follows the simulation on the screen step by step. He can see the token's movement through the GN.

The simulation can be manual or automatic. In manual simulation each GN step may be defined by the user. In automatic simulation, only the first step is defined by the user. Each next step is generated by the system (GN model). In both cases, results of the traversed steps are generated.

The user can define the appearance of the results of the GN model simulation. He can use histogramms and tables for representation of the change in the characteristic selected by him for different tokens. He can follow the relations between different characteristics for one or several tokens. He can follow which transition was not active in a selected time interval, which place was not reached by tokens and other features.

§ 5.6.4. Conclusions

The STGNs are a powerful tool using the GNs for the study of the properties of the practical and theoretical concurrent processes. Their software support and development has to be continued. Currently, the STGNs is used for description and simulation of the GNs based models, described in the this book.

The next step is to make these tools more user-friendly and to extend the class of GNs, which can be described and simulated with them.

REFERENCE:

[1] K. Jensen and R. Shapiro, A tool package, supporting to use of Colored Petri Nets, Petri Net Newsletter, Vol. 35 (1988), 22-36.

Appendix 1: REMARKS ON GENERALIZED NETS

In this Appendix we shall give the basic definitions related with the concept "generalized net" following the contents of the book "Generalized Nets".

*
* *

§ 1. On the concept "generalized net"

We shall give a formal definition of the concepts "transition of a GN" and "a GN".

Every <u>transition</u> is described by a seven-tuple:

$$Z = \langle L', L'', t_1, t_2, r, M, \Box \rangle,$$

where

a) L' and L'' are finite, non-empty <u>sets of places</u> (<u>the transition's input and output places</u> respectively); for the transition in Fig. 1.1 they are $L' = \{l'_1, l'_2, \ldots, l'_m\}$ and $L'' = \{l''_1, l''_2, \ldots, l''_n\}$.

b) t_1 is the <u>current time-moment</u> of the processes's firing;

c) t_2 is the <u>current value of the duration</u> of its activity;

d) r is the <u>transition's condition</u> determining the tokens which shall

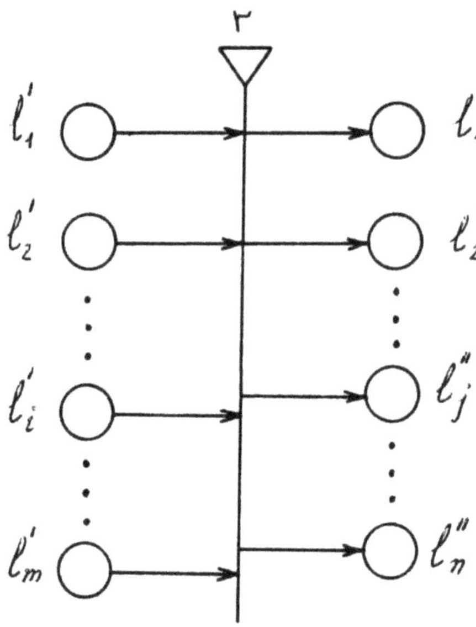

Fig. 1.1.

transfer from the transition's inputs to its outputs; it has the form of an index matrix (see App. 2):

$$r = \begin{array}{c|ccccc} & l''_1 & \ldots & l''_j & \ldots & l''_n \\ \hline l'_1 & & & & & \\ \vdots & & & & & \\ l'_i & & & r_{i,j} & & \\ \vdots & & & (r_{i,j} \text{ - predicates}) & & \\ l'_m & & & (1 \leq i \leq m, \ 1 \leq j \leq n) & & \end{array} ,$$

(i,j) denotes the element which corresponds to the i-th input and j-th output places; these elements are predicates and when the truth value of the (i,j)-th element is true, the token from i-th input place can be transferred to j-th output place; otherwise, it is not possible;

e) M is an index matrix of the capacities of transition's arcs:

$$M = \begin{array}{c|ccccc} & l''_1 & \ldots & l''_j & \ldots & l''_n \\ \hline l'_1 & & & & & \\ \vdots & & & m_{i,j} & & \\ l'_i & & & & & \\ \vdots & & (m_{i,j} \geq 0 \text{ - natural numbers}) & & & \\ l'_m & & (1 \leq i \leq m, 1 \leq j \leq n) & & & \end{array} ;$$

f) □ is an object having a form similar to a Boolean expression. In it the variables are all symbols which mark the transition's inputs nemes, and the Bollean operations "∧" and "v" determine the following conditions:

$\wedge(l_{i_1}, l_{i_2}, \ldots, l_{i_u})$ - every place $l_{i_1}, l_{i_2}, \ldots, l_{i_u}$ must contain at least one token,

$\vee(l_{i_1}, l_{i_2}, \ldots, l_{i_u})$ - in all places $l_{i_1}, l_{i_2}, \ldots, l_{i_u}$ must contain at least one token, where $\{l_{i_1}, l_{i_2}, \ldots, l_{i_u}\} \subset L'$.

The ordered four-tuple

$$E = \langle\langle A, \pi_A, \pi_L, c, f, \theta_1, \theta_2\rangle, \langle K, \pi_K, \theta_K\rangle, \langle T, t^o, t^*\rangle, \langle X, \Phi, b\rangle\rangle$$

is called a Generalized Net (GN), if:

a) A is a set of transitions;

b) π_A is a function giving the priorities of the transitions, i.e., $\pi_A : A \to N$, where $N = \{0, 1, 2, \ldots\} \cup \{\infty\}$;

c) π_L is a function giving the priorities of the places, i.e., $\pi_L : L \to N$, where $L = pr_1 A \cup pr_2 A$, and $pr_i X$ is the i-th projection of the n-dimensional set, where $n \in N$, $n \geq 1$ and $1 \leq k \leq n$ (obviously,

L is the set of all GN-places);

d) c is a function giving the capacities of the places, i.e., $c: L \to N$;

e) f is a function which calculates the truth values of the predicates of the transition's conditions (for the GN described here, let the function f have the value "false" or "true", i.e., a value from the set {0, 1}. In § 3.1, we shall describe other types of nets in which this function will have a value in the interval [0, 1] or in the set [0, 1] x [0, 1]);

f) θ_1 is a function giving the next time-moment when a given transition can be activated, i.e., $\theta_1(t) = t'$, where $t, t' \in [T, T + t^*]$ and $t \leq t'$. The value of this function is calculated at the moment when the transition ends its functioning. Below we shall discuss a special representation of this function;

g) θ_2 is a function giving the duration of the activity of a given transition, i.e., $\theta_2(t) = t'$, where $t \in [T, T + t^*]$ and $t' \geq 0$. The value of this function is calculated at the moment when the transition starts its functioning. It has another representation as the above one;

h) K is the set of the GN's tokens. In some cases, it is convenient to consider this set in the form

$$K = \bigcup_{l \in Q^I} K_l ,$$

where K_l is the set of tokens which enter the net from place l, and Q^I is the set of all net's input places;

i) π_K is a function giving the priorities of the tokens, i.e., $\pi_K : K \to N$;

j) θ_K is a function giving the time-moment when a given token can enter in the net, i.e., $\theta_K(\alpha) = t$, where

$\alpha \in K$,

$t \in [T, T + t^*]$;

k) T is the time-moment when the GN starts functioning. This moment is determined about a fixed (global) time-scale;

l) t^o is an elementary time-step, related to the fixed (global) time-scale;

m) t^* is a duration of the functioning of the net;

n) X is the set of all initial characteristics which can receive the tokens when they enter the net;

o) Φ is a characteristic function which gives new characteristic to every token when it makes a transfer from input to output place of a given transition. As some of the above functions this function can be represented in another form, shown below.

p) b is a function giving the maximum number of characteristics which can receive a given token, i.e., $b: K \rightarrow N$. If for a certain token α, $b(\alpha) = 1$, the token will enter the net with an initial characteristic (as a zero-characteristic). After this, it will receive only the characteristic of the previous. When $b(\alpha) = \infty$, the token α will receive all possible characteristics. When $b(\alpha) = k < \infty$, except its zero-characteristic, the token α will keep the last k as its characteristics (previous characteristics will be "forgotten"). Hence, in general, every token α has $b(\alpha) + 1$ characteristics.

We must note that this definition similarly to the ordinary GN definition is not fully formalized, because if we fully formalize the transition conditions and the characteristic functions of the GNs, a smaller class of GNs will be obtained.

It is convenient to assume that the functions f, θ_1, θ_2 and

Φ have the following forms, also:

$f = \bigcup_{i=1}^{|A|} f_i$, where f_i calculates the truth-values of the predicates of the i-th GN transition (as we shall see in § 3.1, they cannot have values "true" and "false"; they can have fuzzy or intuitionistic fuzzy (see App. 2) values; different transition conditions can be calculated by different ways);

$\theta_1 = \bigcup_{i=1}^{|A|} \theta_1^i$, where θ_1^i calculates the next time-moment of the activation of the i-th GN transition;

$\theta_2 = \bigcup_{i=1}^{|A|} \theta_2^i$, where θ_2^i calculates the duration of the active state of the i-th GN transition;

$\Phi = \bigcup_{i=1}^{|L|} \Phi_i$, where Φ_i calculates the tokens' characteristics, which they will receive in the i-th GN place.

A given GN may not have some of the components. Then the places of such components will be marked by "*". The GNs with such form generate a special class of GNs, which we shall describe in § 2.1.

The static part of a given GN is determined by the elements of the set $pr_{1,2,6,7} A$, where for a given n-dimensional set X ($n \geq 2$)

$$pr_{i_1, i_2, \ldots, i_k} X = \prod_{j=1}^{k} pr_{i_j} X$$

($1 \leq i_j \leq n$, $1 \leq j \leq k$, $i_{j'} \neq i_{j''}$ for $j' \neq j''$), i.e. the static part of a GN is determined by input and output transition places, by index matrix of the arcs and by the type of the transition. The dynamical character of the net comes from the GN's tokens and the transitions conditions ($pr_5 A$), the temporal character comes from the components T, t^{o}, t^{*} and from the elements of the set $pr_{3,4} A$. Finally, the components Φ, X and b play the part of the GN's memory.

The different functions are also related to these four GN components: functions π_{A}, π_{L}, c to the static structure; f, π_{K} to the dynamical elements; θ_{1}, θ_{2} and θ_{K} to the temporal components.

One very important restriction for the GNs is the following. The transition condition predicates cannot be related to future events for the GN.

*
* *

§ 2. Reduced GNs

For two subsets Σ' and Σ'' of Σ - the class of all GNs let us define (see also definitions of the relations $\sqsubset$, $\sqsubset_{*}$, $\approx$, $\approx_{*}$ in § 5:

$\Sigma' \vdash \Sigma''$ <u>iff</u> the functioning and the results of the work of every element of Σ'' can be described by some element of Σ';

$\Sigma' \vdash \Sigma''$ <u>iff</u> $\neg(\Sigma' \vdash \Sigma'')$,

$\Sigma' \dashv\vdash \Sigma''$ <u>iff</u> $(\Sigma' \vdash \Sigma'')$ & $(\Sigma'' \vdash \Sigma')$,

$\Sigma' \dashv\vdash \Sigma''$ <u>iff</u> $\neg(\Sigma' \dashv\vdash \Sigma'')$.

A given GN may not have some of the components. Then the places of such components will be marked by "*". The GNs with such form generate a special class of GNs.

Let us define the set of the GN's components

$$\Omega = \{A, \pi_{A}, \pi_{L}, c, f, \theta_{1}, \theta_{2}, K, \pi_{K}, \theta_{K}, T, t^{o}, t^{*}, X, \Phi, b\} \cup \{A_{i} / 1 \le i \le 7\},$$

where $A_{i} = pr_{i} A$ $(1 \le i \le 7)$, i.e., $A_{i} \in \{L', L'', t_{1}, t_{2}, r, M, \Box\}$ and and let $Y \in \Omega$.

By Σ^{Y} we shall denote the class of those GNs which do not have the Y-component.

Obviously, $\Sigma \vdash \Sigma^{Y}$ for every $Y \in \Omega$. Elements of the class Σ^{Y} we shall call "reduced GNs".

Evidently

$$\Sigma^{A} = \Sigma^{A_1} = \Sigma^{A_2} = \Sigma^{K} = \emptyset,$$

because there does not exist a GN without graphical structure (the component A) and without tokens (the component K); without input and output places of its transitions (components A_1 and A_2 respective).

The class Σ^{Y} will be called "a 1-class of reduced GNs" for every $Y \in \Omega$. If $Y_1, Y_2, \dots, Y_s \in \Omega$ for $s \geq 1$, then $\Sigma^{Y_1, Y_2, \dots, Y_s}$ will be called an "s-class of reduced GNs".

We shall note also that, if $\Sigma^{Y_1, Y_2, \dots, Y_s} \neq \emptyset$, and $Y \notin \{A, A_1, A_2, K\}$, then $\Sigma^{Y_1, Y_2, \dots, Y_s, Y} \neq \emptyset$.

THEOREM 2.2.1: $\Sigma^{A_3, A_4, A_6, A_7, \pi_A, \pi_L, c, \theta_1, \theta_2, \pi_K, \theta_K, T, t^o, t^*, X, b} \mapsto \Sigma$.

*

* *

§ 3. Conservative extensions of the GNs

Eight different types of GN-extensions are defined and for each of them it is proved that it is a conservative extension.

First type intuitionistic fuzzy GNs

In the definition of the concept GN it was noticed that the component f (the function which checks the truth values of the predicates) can give different type values. In the first case they are elements of the set {"false", "true"} (or {0, 1}); in the second - it has values in the interval [0, 1] and then the GN which contains such f-component is called a Fuzzy GN; in the third case - it has values in the set [0, 1] x [0, 1] and the corresponding GN is called Intuitionistic fuzzy GN. The above two nets are denoted in brief by FGN1 and IFGN1 respectively. We shall call these first types GNs "first type FGNs" and "first type IFGNs" (for the concept "intuitionistic fuzzy set" see App. 3).

It is obvious that every FGN1 is a IFGN1. Every GN of this type has the form

$$E = \langle\langle A, \pi_A, \pi_L, c, f, \theta_1, \theta_2\rangle, \langle K, \pi_K, \theta_K\rangle, \langle T, t^o, t^*\rangle, \langle X, \Phi, b\rangle\rangle,$$

where the elements of the set A (the IFGN's transitions) are the same as the GN's transitions. All other components, without the components f and Φ are also the same. The function Φ now gives to every token as current characteristic two values: the first coincides with the token characteristic in the sense of the GN's; the second value is an ordered tuple with real numbers, every one of which is an element of the set [0, 1]. They are equal to the truth values of the predicate of the transition condition between the place, from where starts the transfer of the corresponding token and the place, where this transfer ends. The function f calculates these two values of the corresponding predicate $r_{i,j}$:

$$f(r_{i,j}) = \langle\mu(r_{i,j}), \gamma(r_{i,j})\rangle,$$

where $\mu(r_{i,j})$ is the degree of truth of the predicate $r_{i,j}$, $\gamma(r_{i,j})$ is its degree of false, and (cf. App. 3)

$$\mu(r_{i,j}) + \gamma(r_{i,j}) \leq 1.$$

Second type intuitionistic fuzzy GNs

Here we shall give a complete formal definition of the concept "second type intuitionistic fuzzy GNs" (IFGN2). In this net the tokens are some "quantities" which move inside the net.

The values of the transition condition's predicates can be intuitionistic fuzzings, i.e. they can have degrees of truth and of falseness (see App. 3). The IFGN2 has the form

$$E = \langle\langle A, \pi_A, \pi_L, c, f, \theta_1, \theta_2\rangle, \langle K, \pi_K, \theta_K\rangle, \langle T, t^o, t^*\rangle, \langle X, \Phi\rangle\rangle,$$

where A is the set of the net's transitions which have the ordinary GN-form (cf. § 1) and there exists only one difference: here the index matrix M contains as elements real numbers - capacities of the transition's arcs.

The functions in the first component are similar to those for IFGN1 and they satisfy the same conditions.

The essential difference between IFGN2s and the other GNs is the set K and the functions related to it. Now the elements of K are some "quantities", which have as initial characteristics some "type" (elements of the set X) and which do not receive other characteristics. The function θ_K gives the time-moment when a given token will enter the net as in the ordinary GNs.

The temporal compoments are also as in the other GN-types.

The function Φ has new meaning. Now it is related to the places and they receive by this function characteristics (the quantities of the tokens from the different times in the corresponding places). As in GNs, it can be extended: it can also give other data for the modelled process (e.g. the time-moments for the entry of the "quantities" in the places.

The definition of the second type FGN (FGN2) is analogous.

For IFGN2 the following are valid:

THEOREM 3.2.1: For every IFGN2 there exists a GN which represents it.

THEOREM 3.3.2: For every GN there exists a IFGN2 which represents it.

Colour GNs

They have:

(a) tokens coloured in n different ways ($1 \le n \le \infty$),

(b) arcs coloured in k different ways ($1 \le k \le \infty$).

Arcs of various colours will spring off from every place (but not more than k in number), and also arcs of various colours will be going into every place (but not more than k in number). Thus every GN's transition will have the form of Fig. 3.1, where $1 \le k_i' \le k$ ($1 \le i \le p$) and $1 \le k_j'' \le k$ ($1 \le j \le s$).

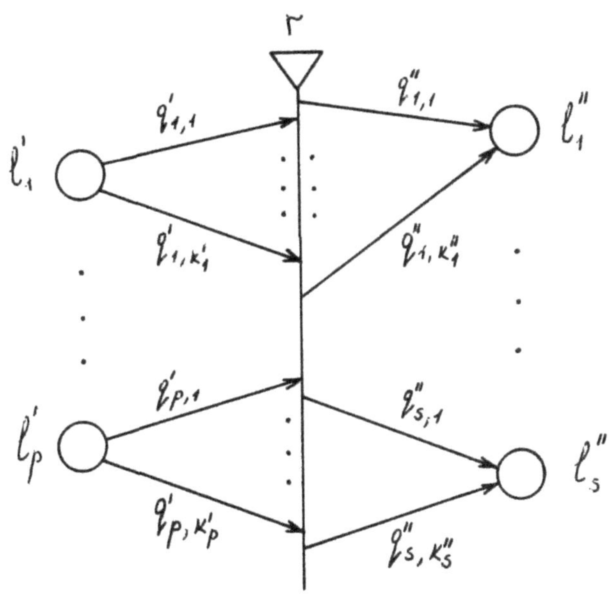

Fig. 3.1.

$(1 \le i \le p)$ and $1 \le k''_j \le k$ $(1 \le j \le s)$.

The token tinted in colour which is marked with number i (in brief "colour i") should only pass through a coloured arc, the number of which is an element of the set I_i. An arc numbered j should host only tokens tinted in the colour with a number which is an element of the set J_j. Obviously:

$$\left| \bigcup_{i=1}^{n} I_i \right| = n ,$$

$$\left| \bigcup_{j=1}^{k} J_j \right| = k .$$

The condition of the above CGN transition (which is an arbitrary one) has the form

$$r = \begin{array}{c|ccccccc} & (l''_1, q''_{1,1}) & \ldots & (l''_1, q''_{1,k''_1}) & \ldots & (l''_s, q''_{s,1}) & \ldots & (l''_s, q''_{s,k''_s}) \\ \hline (l'_1, q'_{1,1}) & & & & & & & \\ \vdots & & & & & & & \\ (l'_1, q'_{1,k'_1}) & & & & r_{q'_{i,\mu}, q''_{j,\gamma}} & & & \\ \vdots & & & & & & & \\ (l'_p, q'_{p,1}) & & & & & & & \\ \vdots & & & (1 \le \mu \le k'_i, & 1 \le \gamma \le k''_j, & 1 \le i \le p, & 1 \le j \le s) & \\ (l'_p, q'_{p,k'_p}) & & & & & & & \end{array}$$

where $r_{q'_{i,\mu}, q''_{j,\gamma}}$ is a predicate related to the $q'_{i,\mu}$-th output arc of the input (for the given transition) place l'_i and to the $q''_{j,\gamma}$ - th input arc of the output place l''_j. Analogously, the M-component of the transition has the form

$$M = \begin{array}{c|c} & (l''_1, q''_{1,1}) \ldots (l''_1, q''_{1,k''_1}) \ldots (l''_s, q''_{s,1}) \ldots (l''_s, q''_{s,k''_s}) \\ \hline (l'_1, q'_{1,1}) & \\ \vdots & \\ (l'_1, q'_{1,k'_1}) & m_{q'_{i,\mu},\, q''_{j,\gamma}} \\ \vdots & \\ (l'_p, q'_{p,1}) & (1 \le \mu \le k'_i,\ 1 \le \gamma \le k''_j,\ 1 \le i \le p,\ 1 \le j \le s) \\ \vdots & \\ (l'_p, q'_{p,k'_p}) & \end{array}$$

where $m_{q'_{i,\mu},\, q''_{j,\gamma}} \ge 0$ are natural numbers.

The net so constructed has a set of tokens K of the form $K = \bigcup_{i=1}^{n} K_i$, where K_i is the set of tokens which are coloured in i-th colour. Obviously, every ordinary GN is a CGN for which $n = K = 1$.

GNs with interval time for activation

The transition firing time-moment of this extension of the GNs is substituted by an interval. In that way the transitions in the new net acquire the form:

$$\bar{Z} = \langle L', L'', \partial, t_2, r, M, \Box \rangle,$$

where $\partial = [t', t'']$ and $T \le t' \le t'' \le T + t^*$. When the tokens in the input places of $\bar{Z}$ are already enough to satisfy the transition type $\Box$ and TIME $\in \partial$, the transition $\bar{Z}$ will be fired.

Let the GN $\bar{E}$ be a net which has at least one transition of the form described above. For this the function θ_1 is extended to function

$\bar{\theta}_1$, which to every time-interval $\theta = [t'_1, t''_1]$ juxtaposes a new time-interval $[t'_2, t''_2]$ for which $(T \leq)\ t''_1 \leq t'_2 \leq t''_2 \leq T + t^*$, i.e., the next time-interval for firing of the transition $\bar{Z}$ will come not earlier than the moment when this transition ends its former firing.

GNs with a complex structure

The transition forms of GNs are graphically similar to those of the E-nets. For the Petri nets the graphical structure is not the same as for GNs. It is proved that if a constructed GN which allows more than one arc entering each of its places and more than one arc coming out of each of its places, then this extension of the GNs will also be conservative and there it is called "a GN with a complex structure".

GNs with global memory

In the modelling of real processes it is appropriate to keep data (while the GN is functioning) or to determine the values of different parameters related to these processes. Thus, the definition of the concept GN can be extended with the addition of a new component B, which we shall call "global memory". The component B can be seen as a list of such functions that will change their values in the process of GN functioning. Let a GN with component B be called "a GN with global memory" (GNGM). The formal definition of this object is:

$$E^* = \langle\langle A, \pi_A, \pi_L, c, f, \theta_1, \theta_2\rangle, \langle K, \pi_K, \theta_K\rangle, \langle T, t^o, t^*\rangle,$$
$$\langle B, X, \Phi, b\rangle\rangle,$$

where all components are as in § 1.1 and the component B is as described above.

GNs with optimization components

Some GN's models correspond to real processes whose functions do not stop and which determine standard (given before) solutions for the process's continuation. For these GNs it is necessary to add such components that secure their non-stop functioning. For this aim, we shall construct a new extension of the ordinary GNs - GNs with additional clocks (GNACs). They have transitions with the form:

$$Z = \langle L', L'', t_1, t_2, r, t_3, \bar{r}', M, \square \rangle,$$

where the components common with the ordinary transitions are the same and

a) $\bar{r}'$ is a (0, 1)-index matrix with the form

$$\bar{r}' = \begin{array}{c|ccccc} & l''_1 & \dots & l''_j & \dots & l''_n \\ \hline l'_1 & & & & & \\ \vdots & & & \bar{r}'_{i,j} & & \\ l'_i & & & & & \\ \vdots & & & (r_{i,j} \in \{0, 1\}) & & \\ l'_m & & & (1 \le i \le m, 1 \le j \le n) & & \end{array},$$

where $L' = \{l'_1, l'_2, \dots, l'_m\}$ and $L'' = \{l''_1, l''_2, \dots, l''_n\}$,

b) t_3 is the maximal time-duration for a check of the truth of the predicates of the transition condition r. In the case that time is increased the checking of the predicates of r shall be reduced and their respective values from the matrix $\bar{r}'$ will be assigned.

Obviously, $t_3 \le t_2$, but if $t_3 > t_2$ the process of checking of the predicate's truth can be stopped before the time necessary for this time-moment.

If A is the set of the GNAC's transitions the GNAC has the form

$$E = \langle\langle A,\ \pi_A,\ \pi_L,\ c,\ f,\ \theta_1,\ \theta_2,\ \theta_3,\ \sigma\rangle,\ \langle K,\ \pi_K,\ \theta_K\rangle,\ \langle T,\ t^o,\ t^*\rangle,$$

$$\langle X,\ \Phi,\ \Phi',\ b\rangle\rangle,$$

where

a) θ_3 is a function giving the duration of the transition's condition checking, i.e., $\theta_3(t) = t'$, with $t \in [T, T + t^*]$ and $t' \geq 0$. The value of this function is calculated at the moment when the transition starts its functioning and, after this, at every time-moment, when the condition's predicates are calculated and the current step of the token's transfers are finished and there exists more such steps. This function determines the values of the transition component t_3, and if $\theta_3(TIME) = t'$ and $TIME + t' > t_1 + t_2$, the component t_3 does not receive new value.

b) σ is a function which to the predicates of r juxtaposes (0, 1)-values by an advance determined way. Its calculations are made immediatelly after each checking of the values of the function θ_3.

c) Φ' is a function which determines the new characteristics of the tokens which enter the transition's output places after applying the specifics of the new nets mechanism.

*
* *

§ 4. GNs and other objects

In this chapter, it is proved that the functioning of the results of the work of the different types of modifications of the Petri nets and of the finite automata and the Turing machines can be represented by GNs. A GN which makes the same for each ordinary Petri net is constructed.

* * *

§ 5. Algebraic aspect of the theory of GNs

For the two transitions Z_1 and Z_2 we shall define

$Z_1 = Z_2$ iff $(\forall i: 1 \le i \le 7)(pr_i Z_1 = pr_i Z_2)$;

$Z_1 \subset Z_2$ iff $(\forall i: 1 \le i \le 2)(pr_i Z_1 \subset pr_i Z_2)$ &

$(\forall i: 3 \le i \le 4)(pr_i Z_1 = pr_i Z_2)$ &

$(\forall i: 5 \le i \le 6)(pr_i Z_1 \subset_1 pr_i Z_2)$ & $(pr_i Z_1 \subset_2 pr_i Z_2)$,

where

$\subset_1$ is a relation of inclusion over index matrices and if

$A = [K_1, L_1, \{a_{i,j}\}]$, $B = [K_2, L_2, \{b_{i,j}\}]$, then

$A \subset_1 B$ iff $(K_1 \subset K_2)$ & $(L_1 \subset L_2)$ & $(\forall i \in K_1)(\forall j \in L_1)(a_{i,j} = b_{i,j})$

$\subset_2$ is a relation of inclusion over Boolean expressions and, for two of these expressions a and b

$a \subset_2 b$ iff it is getting the expression a after removing the arguments of b, which are not arguments of a and the logical operations, associated to them.

Over the transitions $Z_i = \langle L_1^i, L_2^i, t_1^i, t_2^i, r^i, M^i, \square^i \rangle$ $(i = 1, 2)$, we shall define three operations. It is necessary for the following to be valid: if place $l \in pr_1 Z_i \cap pr_2 Z_i$ and $l \in pr_s Z_{3-i}$, then $l \in pr_{3-s} Z_{3-i}$ for $1 \le i \le 2$, $1 \le s \le 2$. These operations are

a) a <u>union</u> (the necessary conditions for this operation are $t_j^1 = t_j^2$ $(j = 1, 2)$, and (as if $l \in pr_s Z_i$, this is not possible), that $l \in pr_{3-s} Z_{3-i}$ for $1 \le i \le 2$, $1 \le s \le 2$):

$$Z_1 \cup Z_2 = \langle L_1^1 \cup L_1^2, L_2^1 \cup L_2^2, t_1^1, t_2^1, r^1 + r^2, M^1 + M^2, \vee(\square^1, \square^2)\rangle,$$

b) an <u>intersection</u> (with the above conditions):

$$Z_1 \cap Z_2 = \langle L_1^1 \cap L_1^2, L_2^1 \cap L_2^2, t_1^1, t_2^1, r^1 \times r^2, M^1 \times M^2, \wedge(\square^1, \square^2)\rangle.$$

c) a <u>composition</u> (with the above condition and with the condition $L_1^1 \cap L_1^2 = L_2^1 \cap L_2^2 = \emptyset$):

$$Z_1 \circ Z_2 = \langle L_1^1 \cup (L_1^2 - L_2^1), L_2^2 \cup (L_2^1 - L_1^2), t_1^1, t_2^1, r^1 \circ r^2, M^1 \circ M^2, \vee(\square^1, \bar{\square}^2)\rangle,$$

where $\bar{\square}$ results from $\square$ after removing all its arguments, whose identifications are elements of the set $L_2^1 \cap L_1^2$.

It is possible that $L_1^1 \cap L_1^2 = \emptyset$ and $L_2^1 \cap L_2^2 = \emptyset$. In this case $Z_1 \cap Z_2 = Z^o$, where Z^o is the empty transition (for some other components as M, r, $\square$ will be degenerated).

The operations described below are unique in the Petri net theory. They can be transfered on practically all other types of Petri nets (obviously with some modifications related to the structure of the corresponding nets). These operations are very useful for constructing GN models of real processes.

Before introducing the different GN's operations, we shall formulate some conditions which the arguments of these operations (different GNs) must satisfy.

We shall assume that if a place takes part simultaneously in two nets, then in both locations the place has

(a) equal capacities,

(b) the same characteristic functions

(c) equal possibility to accept tokens from K, i.e. if in the first net (which is a model of some process) the tokens which are elements of the set $K' \subset K$ could pass through the examined place, while in the second net tokens which are elements of the set $K'' \subset K$ could pass through this place, then $K' = K''$,

(d) equal values of the priority functions,

(e) one and the same numeration/notation.

Similarly, we shall expect that if two places are respectively an input one and an output one for two transitions in both nets, then

(a) the capacities of the connecting arcs are equal,

(b) the time for passing from the input to the output place should be one and the same,

(c) they should be in the same consequence, i.e. the place which is an input (output) one in the first of the transitions should be the same in the other transition too.

Let E_1 and E_2 be two GNs and let for $i = 1, 2$:

$$E_i = \langle\langle A_i, \pi^i_A, \pi^i_L, c^i, f^i, \theta^i_1, \theta^i_2\rangle, \langle K_i, \pi^i_K, \theta^i_K\rangle, \langle T_i, t^o_i, t^*_i\rangle,$$

$$\langle X_i, \Phi_i, b_i\rangle\rangle .$$

A <u>union</u> will be called the object:

$$E_1 \cup E_2 = \langle\langle A_1 \,\overline{\cup}\, A_2, \pi^1_A \cup \pi^2_A, \pi^1_L \cup \pi^2_L, c^1 \cup c^2, f^1 \cup f^2, \theta^1_1 \cup \theta^2_1,$$

$$\theta^1_2 \cup \theta^2_2\rangle, \langle K_1 \cup K_2, \pi^1_K \cup \pi^2_K, \theta^1_K \cup \theta^2_K\rangle, \langle \min(T_1, T_2), GCD(t^o_1, t^o_2),$$

$$\max_{1\le i\le 2}(T_i + t^*_i . t^o_i/GCD(t^o_1, t^o_2) - \min(T_1, T_2))>,$$

$$<X_1 \cup X_2, \Phi_1 \cup \Phi_2, b_1 \cup b_2>>,$$

where

$$A_1 \bar{\cup} A_2 = \bigcup_{i=1}^{2} \{Z/(Z \in A_i) \& (\forall Z' \in A_{3-i})(Z \cap Z' = Z^o)\} \cup$$

$$\bigcup_{i=1}^{2} \{Z/(\exists Z' \in A_i)(\exists Z'' \in A_{3-i})(Z' \cap Z'' \neq Z^o) \& (Z = Z' \cup Z'')\}.$$

A <u>composition</u> of the above nets will be called the object:

$$E_1 \circ E_2 = \begin{cases} E_1, & \text{if } T_2 + t^*_2 < T_1 \\ E_3, & \text{if } T_1 \le T_2 + t^*_2 \end{cases},$$

where

$$E_3 = <<A_1 \bar{\cup} A_2, \pi^1_A \cup \pi^2_A, \pi^1_L \cup \pi^2_L, c^1 \cup c^2, f^1 \cup f^2, \theta^1_1 \cup \theta^2_1, \theta^1_2 \cup \theta^2_2>,$$

$$<K_1 \cup K_2, \pi^1_K \cup \pi^2_K, \theta^1_K \cup \theta^2_K>, <T_1, GCD(t^o_1, t^o_2), \max_{1\le i\le 2}(T_i +$$

$$t^*_i . t^o_i/GCD(t^o_1, t^o_2) - T_1)>, <X_1 \cup X_2, \Phi_1 \cup \Phi_2, b_1 \cup b_2>>.$$

Before introducing the next operation over GNs, we shall defiene a local operator O - an <u>operators for place identification</u>, which juxtaposes to every given GN E and to every given places $l_1 \in Q^I_E$ and $l_2 \in Q^O_E$ a new GN

$$O(E, l_1, l_2) = <<(A - \{<L', L'', t_1, t_2, r, M, \square>/l'' \in L''\}) \cup$$

$$\{<L', \bar{L}'', t_1, t_2, r, M, \square>/(l'' \in L'') \& (\bar{L}'' = (L'' - \{l''\}) \cup \{l'\})$$

$\& (\langle L', L'', t_1, t_2, r, M, \square\rangle \in A)\}\rangle, \pi_A, \pi_L, c, f, \theta_1, \theta_2\rangle, \langle K,$

$\pi_K, \theta_K\rangle, \langle T, t^o, t^*\rangle, \langle X, \Phi, b\rangle\rangle.$

It can be easily seen, that the sets

$$\{\langle L', L'', t_1, t_2, r, M, \square\rangle / l'' \in L''\}$$

and

$$\{\langle L', \bar{L}'', t_1, t_2, r, M, \square\rangle / (l'' \in L'') \& (\bar{L}'' = (L'' - \{l''\}) \cup \{l'\} \&$$
$$(\langle L', L'', t_1, t_2, r, M, \square\rangle \in A)\}$$

are one-element sets (of transitions).

We shall, extend the operator O over two subsets of L, as follows. If $L' = \{l'_1, l'_2, \dots, l'_n\} \subset Q^I_E$ and $L'' = \{l''_1, l''_2, \dots, l''_n\} \subset Q^O_E$ and $l'_i \neq l'_j$, $l'_i \neq l''_j$, $l''_i \neq l'_j$ $(1 \leq i, j \leq n)$, then

$$O(E, L', L'') = O(\dots O(O(E, l'_1, l''_1), l'_2, l''_2)\dots, l'_n, l''_n).$$

Obviously, the new object will be a GN too.

The object E* for which there exist sets $L' = \{l'_1, l'_2, \dots, l'_n\} \subset Q^I_E$ and $L'' = \{l''_1, l''_2, \dots, l''_n\} \subset Q^O_E$ and $E* = O(E, L', L'')$ for some fixed GN E for which $L = L_E$, and for every i $(1 \leq i \leq n)$ there exists an oriented path from a place l'_i to a place l''_i in E, we shall call <u>iteration</u> of the GN E.

For every two GNs E_1 and E_2 different types of relations are defined in [*], but we shall give here only these which are used in the book.

Let us define for a given GN E two sets, one of tokens K and one of the initial token's characteristics X:

$$K(E) = \{\alpha / (\alpha \in K) \ \& \ (\theta_K(\alpha) < T + t^*)\},$$

$$X(E) = \{x_o^\alpha / \alpha \in K(E)\},$$

and let $X(\alpha)$ be the set of all different characteristics, with which the token α can start its transfer in the GN E. Obviously,

$$X(E) \subset \bigcup_{\alpha \in K} X(\alpha).$$

For a given token $\alpha \in K$ and a given initial characteristic $x \in X(E)$ let

$$E(\alpha, x) = \begin{cases} \langle \alpha, x_{fin}^\alpha \rangle, & \text{if } \alpha \in K(E) \text{ and } x \in X(\alpha) \\ \langle \alpha, x \rangle, & \text{otherwise} \end{cases},$$

$$E\{\alpha, x\} = \begin{cases} \langle \alpha, x, x_1^\alpha, x_2^\alpha, \ldots, x_{fin}^\alpha \rangle, & \text{if } \alpha \in K(E) \text{ and } x \in X(\alpha) \\ \langle \alpha, x \rangle, & \text{otherwise} \end{cases},$$

where x_{fin}^α is the final characteristic of the token α in the GN E.

Let us define

$$E_\emptyset(\alpha, x) = E_\emptyset\{\alpha, x\} = \langle \alpha, x \rangle.$$

Below we shall introduce four relations between the GNs:

$E_1 \sqsubset E_2$ <u>iff</u> $(K(E_1) \subset K(E_2))$ & $(X(E_1) \subset X(E_2))$ & $(\forall \alpha \in K(E_1))$
$(\forall x \in X(E_1))(E_1(\alpha, x) = E_2(\alpha, x))$

$E_1 \sqsubset_* E_2$ <u>iff</u> $(K(E_1) \subset K(E_2))$ & $(X(E_1) \subset X(E_2))$ & $(\forall \alpha \in K(E_1))$
$(\forall x \in X(E_1))(\exists i_1, i_2, \ldots, i_s : 1 \le i_1 < \ldots < i_s \le fin)$
$(E_1\{\alpha, x\} = pr_{i_1, i_2, \ldots, i_s} E_2\{\alpha, x\}),$

$E_1 \approx E_2$ <u>iff</u> $(E_1 \sqsubset E_2)$ & $(E_2 \sqsubset E_1)$,

$E_1 \approx_* E_2$ iff $(E_1 \sqsubset_* E_2)$ & $(E_2 \sqsubset_* E_1)$.

Obviously, the relations $\sqsubset_*$ and $\approx_*$ are stronger than the relations $\sqsubset$ and $\approx$ respectively.

The first GN-definition (in § 1.) in some sense is a deductive one. Below we shall introduce a second GN-definition. In the above sense, the new definition will be of an inductive type.

Let Z be a given object, having the grapical structure in Fig. 1.1 and the following components:

a) L′ - a set of places which we shall call input places,

b) L″ - a set of places which we shall call output places,

c) t_1 - a time-moment about some fixed time-scale (with an elementary time-step t^o),

d) t_2 - a real number which corresponds to the length of a time-interval in the above mentioned time-scale,

e) r - an index matrix (see App. 1) having the form

$$r = \begin{array}{c|c} & l''_1 \; \ldots \; l''_j \; \ldots \; l''_n \\ \hline l'_1 & \\ \vdots & r_{i,j} \\ l'_i & \\ \vdots & (r_{i,j} \text{ - predirates}) \\ l'_m & (1 \le i \le m,\ 1 \le j \le n) \end{array} ,$$

(i,j)-th element of which correspond to the i-th input and j-th output places, these elements being predicates,

f) M - an index matrix having the form

$$M = \begin{array}{c|c} & l''_1 \; \dots \; l''_j \; \dots \; l''_n \\ \hline l'_1 & \\ \vdots & m_{i,j} \\ l'_i & \\ \vdots & (m_{i,j} \geq 0 \text{ - natural numbers}) \\ l'_m & (1 \leq i \leq m,\ 1 \leq j \leq n) \end{array};$$

g) □ is an object having a form similar to an Boolean expression. Its variables are exactly the symbols which mark the Z's input places' names; the Boolean operations "∧" and "∨" determine the analogous conditions from § 1 and

1) $\wedge(l_{i_1}, l_{i_2}, \dots, l_{i_u})$ - every one of the places $l_{i_1}, l_{i_2}, \dots, l_{i_u}$ must contain at least one token,

2) $\vee(l_{i_1}, l_{i_2}, \dots, l_{i_u})$ - all the places $l_{i_1}, l_{i_2}, \dots, l_{i_u}$ must contain at least one token, where $\{l_{i_1}, l_{i_2}, \dots, l_{i_u}\} \subset L'$.

The object described above will be called a <u>transition</u>.

The <u>inductive definition of the concept GN</u> is as follows:

(1) The object which has the form of a transition is called a <u>GN</u>, if to it are added

a) π_A - a function giving a natural number, being the priority of the transition,

b) π_L - a function giving the priorities of the transition's places,

c) c - a function giving the capacities of the transition's places,

d) f - a function which calculates the truth values of the predicates of the index matrix r;

e) θ_1 - a function giving the next time-moment when the given transition can be activated, i.e. $\theta_1(t) = t'$, where $t, t' \in [T, T+t^*]$ and $t \leq t'$. The value of this function is calculated at the moment when the transition ends its functioning;

f) θ_2 - a function giving the duration of the activity of a given transition, i.e., $\theta_2(t) = t'$, where $t \in [T, T + t^*]$ and $t' \geq 0$. The value of this function is calculated at the moment when the transition starts its functioning;

g) K - a set of tokens;

h) π_K - a function giving the priorities of the tokens ($\pi_K : K \to N$);

i) θ_K - a function giving the time-moment when a given token can enter the net, i.e. $\theta_K(\alpha) = t$, where

$\alpha \in K$,

$t \in [T, T + t^*]$;

j) T - a time-moment when the GN starts functioning. This moment is determined on a fixed time-scale and the first value of $t_1 = T$;

k) t^o - an elementary time-step related to the fixed time-scale;

l) t^* - a duration of the functioning of the net;

m) X - a set of all initial characteristics which can receive the tokens when they enter the net;

n) Φ - a characteristic function which gives new characteristic to every token when it makes a transfer from input to output place of a given transition;

o) b - a function giving the maximal number of characteristics which can receive a given token (i.e. $b: K \to N$) transfering in the net.

(2) If E, E_1 and E_2 are GNs, then

a) $E_1 \cup E_2$ is a GN,

b) $E_1 \circ E_2$ is a GN,

c) $E*$ is a GN.

THEOREM 5.3.1: The two definitions of the concept "GN" are equivalent.

* * *

§ 6. Topological aspect of the theory of GNs

The graphical structure of a given GN is a two-coloured graph. Thus the study of such a structure and checking of the properties of a GN following from these structures are important for the theory of the GNs. Some operators which assign to every GN one- or two-coloured graph and some other operators, which juxtapose to every GN one natural number corresponding to the complexity of the net with respect to a certain criterion, are defined. The class Σ is ordered by relations, generated from the above two types of operators, and some topological properties of Σ are studied.

* * *

§ 7. Logical aspect of the theory of GNs

Logical operators which are similar to the modal operators "necessity" and "possibility" are defined and some of their basic properties are given.

* * *

§ 8. Operator aspect of the theory of GNs

The operator aspect has an important place in the theory of the GNs. Six types of operators are defined in its framework. Every operator assigns to a given GN a new GN, which has some preliminaryly, introduced properties. Particular types of operators are: global (one which transforms globally some GN's components), local (one which transforms some transition components of a given GN), hierarchical (one which substitutes a place or a transition for some part of a given GN, or makes reversal), reducing (one which removes some of the GN components), extending (one which transforms a given GN to the corresponding extending GN) and dynamical (one which determines the different strategies for the tokens' transfer in the net). The basic task of this aspect of the theory of the GNs is a research on the connections and influence between the separate operators.

*
* *

§ 9. Other extensions of the GNs

By analogy with the universal Turing machine, the object Universal GN (UGN) as a GN E^* for which the following is valid:

$$(\forall E \in \Sigma)(E \sqsubset E^*),$$

is intriduced. It is noted that at the moment of writing of [*] (Nov. 1990), the question for the existence of an UGN is open.

Let Ω be a some fixed time-scale. For $T' \in \Omega$ we define

$$\Sigma_{T'} = \{E \,/\, E \in \Sigma \;\&\; pr_1 pr_3 E + pr_3 pr_3 E \leq T'\}.$$

We shall call the GN E^* for which the following is valid,

$$(\forall E \in \Sigma_{T'})(E \sqsubset_* E^*),$$

a T'-UGN about the time-scale Ω. Obviously, the global time-compoments of

$$E = \langle\langle A, \pi_A, \pi_L, c, f, \theta_1, \theta_2\rangle, \langle K, \pi_K, \theta_K\rangle, \langle T, t^o, t^*\rangle, \langle X, \Phi, b\rangle\rangle,$$

satisfy the inequality $T + t^* \leq T'$.

THEOREM 9.1.1: For every time-moment $T' \in \Omega$ about the fixed time-scale

Ω, there exists a T'-UGN E^* for the class $\Sigma_{T'}$.

Let

$$\Sigma_* = \{E \mid E \in \Sigma \ \& \ pr_1 pr_3 E = * \ \& \ pr_3 pr_3 E < \infty\},$$

i.e., every GN of Σ_* has finite time for its functioning. Let E_* be a GN similar to E^* but without the first global temporal component. Then we shall define GN E_* as a universal GN (UGN) for the class Σ_*, i.e.,

$$(\forall E \in \Sigma_*)(E \sqsubset_* E^*).$$

<u>THEOREM 9.1.2</u>: There exists a UGN for the class Σ_*.

Let E be some GN. Let K be a set of tokens of E, and X be a set of initial characteristics for these tokens of K. The object "<u>Self-Modifying GN</u>" (<u>SMGN</u>) is constructed.

Let K^* be a set of tokens which will be called "control tokens". The defining of this set is possible in the framework of the GN theory. The tokens of K^* can be interpreted as GN-tokens with initial characteristics (elements of the set X^*) which contain the identifier "control token" (this is also possible). The second addition in the frame of a given GN E is that the characteristic functions of some places (in which there will be transferred control tokens) give as characteristic identifiers and parameters of some types of operators, defined over the GN. This is also not in contradiction with the definition of GNs. Finally (really, the new): when a certain control token gets as characteristic the identifier of some operator with parameters former or current characteristics of this or another control token, then at this time-moment over the GN E the corresponding operator will be realized.

Obviously, not-every operator can be applied to the net in the time-interval when it functions. Thus, we can define the SMGN as a GN, for which there exists a mechanism μ for realization of the operators over it. This mechanism functions in the following way: at the time-moment when a certain control token of a given GN gets as a characteristic the identifier of a certain (added) operator and the values of its parameters, the mechanism μ realizes this operator.

The basic aims in [*] are:

1. to show the operators which can be realized by μ;
2. to show the relations between GNs and SMGNs.

If Σ_{SMGN} is the class of all SMGNs, then obviously $\Sigma \subset \Sigma_{SMGN}$ and $\Sigma_{SMGN} \vdash \Sigma$, because every GN is a SMGN with an empty set of control tokens.

<u>THEOREM 9.2.1</u>: For every SMGN, there exists a GN which represents it.

The SMGNs are the basic aim of the operator aspect of the theory of GNs.

*
* *

§ 10. Methodological aspect of the theory of GNs

The basic ways for constructing of a GN of a given real process and some ways for using and modifying of already constructed GN-models are discussed.

*
* *

§ 11. Open problems

There are many unsolved or unformulated problems in the theory of GNs. Here we shall introduce some of these problems which are related to the text of the book.

5. To research the different classes of reduced GNs without two or more components, and to show the relations between these classes.

7. To prove that the constructed reduced GN in § 2.2 from the book on GNs (from § 1.1.2) has the least possible number of different GN's components, or to give a counter example of this.
8. What other conservative extensions of the GNs can be defined?
9. To construct GNs which are universal for the classes (sets) of the different Petri net modifications, by analogy with the GNs from § 4.3.
12. To construct an algorithm which for every set of transition graphical structures (obviously they must satisfy some conditions) and for every given GN, construct a GN for which:
    - all its transitions have the forms, which are elements of the given set,
    - both nets have equal tokens with equal initial characteristics and as a result of their functioning equal tokens receive equal final characteristics.
19. To prove or disprove the existence of a universal GN for Σ.
20. To construct other extensions of the GNs and to prove or disprove their conservativness.
21. Chapter 10 is a superficial attempt to give the basics of the methodology of work with GNs. To continue and enrich the research in this area.
22. A list of the basic applications to this moment (Nov. 1990) is given in Chapter 0. To extend the area of the GN's applications.

Appendix 2: GENERALIZED INDEX MATRICES

A definition of "generalized index matrix" (GIM) or "index matrix" is given in [1-3]. Here we shall introduce only the results which are related to this basic text.

Let I be a fixed index set and R be the set of all real numbers. A GIM with generalized index sets K and L $(K, L \subset I)$ will be called the object:

$$\begin{array}{c|cccc} & l_1 & l_2 & \dots & l_n \\ \hline k_1 & a_{k_1,l_1} & a_{k_1,l_2} & \dots & a_{k_1,l_n} \\ k_2 & a_{k_2,l_1} & a_{k_2,l_2} & \dots & a_{k_2,l_n} \\ \vdots & \vdots & \vdots & & \vdots \\ k_m & a_{k_m,l_1} & a_{k_m,l_2} & & a_{k_m,l_n} \end{array} \quad (= [K, L, \{a_{k_i,l_j}\}]),$$

where $K = \{k_1, k_2, \dots, k_m\}$, $L = \{l_1, l_2, \dots, l_n\}$, for $1 \le i \le m$, and for $1 \le j \le n$: $a_{k_i,l_j} \in R$. Let $M_{K,L}$ be the set of all GIMs with index sets K and L, and let $M = \bigcup_{K,L \subset I} M_{K,L}$.

In [3], for the GIMs $A = [K, L, \{a_{k_i, l_j}\}]$, $B = [P, Q, \{b_{p_r, q_s}\}]$ are defined ordinary matrix operations "addition" and "multiplication", and also the following operations

$A + B = [K \cup P, L \cup Q, \{c_{t_u, v_w}\}]$, where

$$c_{t_u, v_w} = \begin{cases} a_{k_i, l_j} &, \text{if } t_u = k_i \in K \text{ and } v_w = l_j \in L - Q \text{ or} \\ & \quad t_u = k_i \in K - P \text{ and } v_w = l_j \in L \\ b_{p_r, q_s} &, \text{if } t_u = p_r \in P \text{ and } v_w = q_s \in Q - L \text{ or} \\ & \quad t_u = p_r \in P - K \text{ and } v_w = q_s \in Q \\ a_{k_i, l_j} + b_{p_r, q_s} &, \text{if } t_u = k_i = p_r \in K \cap P \text{ and } v_w = l_j = q_s \\ & \quad \in L \cap Q \\ 0 &, \text{otherwise} \end{cases};$$

$A \times B = [K \cap P, L \cap Q, \{c_{t_u, v_w}\}]$, where

$$c_{t_u, v_w} = a_{k_i, l_j} \cdot b_{p_r, q_s}, \text{ for } t_u = k_i = p_r \in K \cap P \text{ and } v_w = l_j = q_s \in L \cap Q;$$

$A \,.\, B = [K \cup (P-L), Q \cup (L-P), \{c_{t_u, v_w}\}]$, where

$$c_{t_u, v_w} = \begin{cases} a_{k_i, l_j} &, \text{if } t_u = k_i \in K \text{ and } v_w = l_j \in L - P \\ b_{p_r, q_s} &, \text{if } t_u = p_r \in P - L \text{ and } v_w = q_s \in Q \\ \sum\limits_{l_j = p_r \in L \cap P} a_{k_i, l_j} \cdot b_{p_r, q_s} &, \text{if } t_u = k_i \in K \text{ and } v_w = q_s \in Q \\ 0 &, \text{otherwise} \end{cases};$$

The basic properties of the set M with the above operations are studied in [3]. There some GIMs measures are introduced (for GIMs determinants do not exist).

The given mathematical apparatus may be applied to the GIMs with elements from the sets {0, 1}, [0, 1], or from the class of all predicates, etc. In the first two cases, the operations "+" and "." in R will be substituted by "max" and "min" respectively, and in the third case - by the operations "v" and "∧".

REFERENCES:

[1] Atanassov K. Conditions in Generalized nets, Proc. of the XIII Spring Conf. of the Union of Bulg. Math., Sunny Beach, April 1984, 219-226.

[2] Atanassov K. Dynamical elements in the Generalized nets, AMSE Review Vol. 1 (1985), No. 4, 1-9.

[3] Atanassov K. Generalized index matrices, Comptes rendus de l'Academie Bulgare des Sciences, Vol. 40, 1987, No. 11, 15-18.

Appendix 3: INTUITIONISTIC FUZZY SETS

Following [1,2] we shall give short remarks on the intuitionistic fuzzy sets (IFSs) and other intuitionistic fuzzy objects related to them.

Let a set E be fixed. An IFS A^* in E is an object having the form

$$A^* = \{<x, \mu_A(x), \gamma_A(x)>/x \in E\},$$

where the functions $\mu_A(x): E \rightarrow [0, 1]$ and $\gamma_A(x): E \rightarrow [0, 1]$ define the degree of membership and the degree of non-membership of the element $x \in E$ to the set A, which is a subset of E, respectively, and for every $x \in E$

$$0 \le \mu_A(x) + \gamma_A(x) \le 1.$$

Obviously, every ordinary fuzzy set has the form

$$\{<x, \mu_A(x), 1-\mu_A(x)>/x \in E\}.$$

For simplicity, below we shall write A instead of A^*.

For every two IFSs A and B, the following relations and operations are valid:

$A \subset B$ <u>iff</u> $(\forall x \in E)(\mu_A(x) \le \mu_B(x)\ \&\ \gamma_A(x) \ge \gamma_B(x))$;

$A \supset B$ <u>iff</u> $B \subset A$;

$A = B$ iff $(\forall x \in E)(\mu_A(x) = \mu_B(x)\ \&\ \gamma_A(x) = \gamma_B(x))$

$\bar{A} = \{\langle x, \gamma_A(x), \mu_A(x)\rangle / x \in E\}$;

$A \cap B = \{\langle x, \min(\mu_A(x), \mu_B(x)), \max(\gamma_A(x), \gamma_B(x))\rangle / x \in E\}$;

$A \cup B = \{\langle x, \max(\mu_A(x), \mu_B(x)), \min(\gamma_A(x), \gamma_B(x))\rangle / x \in E\}$;

$A + B = \{\langle x, \mu_A(x) + \mu_B(x) - \mu_A(x).\mu_B(x), \gamma_A(x).\gamma_B(x)\rangle / x \in E\}$;

$A . B = \{\langle x, \mu_A(x).\mu_B(x), \gamma_A(x) + \gamma_B(x) - \gamma_A(x).\gamma_B(x)\rangle / x \in E\}$;

$A - B = A \cap \bar{B}$.

The concept IFS is the basis for the definition of the concepts Intuitionistic Fuzzy Propositional Calculus (IFPC) and such extensions as IF predicative calculus, IF modal logic, etc. (see [3-5]).

To each proposition (in the classical meaning) one can assign its truth value: truth denoted by 1, or falsum by 0. In the case of fuzzy logics, this truth value is a real number in the interval [0, 1] and can be called "truth degree" of a particular proposition. In the case of IFPC is added one more value - "falsum degree" - which will be in the interval [0, 1] as well. Thus one assigns to the proposition p two real numbers $\mu(p)$ and $\gamma(p)$; moreover the following constraint is valid:

$$\mu(p) + \gamma(p) \leq 1.$$

Let this be done by an evaluation function V defined so that

$$V(p) = \langle \mu(p), \gamma(p)\rangle.$$

Hence the function $V: S \to [0, 1] \times [0, 1]$ gives the truth and falsum degrees from the class of all propositions and it can be defined so that it assigns to the logical truth T:

$$V(T) = \langle 1, 0\rangle,$$

and to the logical falsum F:

$$V(F) = \langle 0, 1\rangle.$$

The negation $\neg p$ of the proposition p will be defined through

$$V(\neg p) = \langle\gamma(p),\ \mu(p)\rangle.$$

When

$$\gamma(p) = 1 - \mu(p),$$

i. e.

$$V(p) = \langle\mu(p),\ 1 - \mu(p)\rangle,$$

for $\neg p$ we get

$$V(\neg p) = \langle 1 - \mu(p),\ \mu(p)\rangle,$$

which coincides with the result, e. g., from [6].

Depending on the way of defining, of the operation "$\supset$" can be obtained by different variants of IFPC. One of them is as follows:

$$V(p\ \&\ q) = \langle\min(\mu(p),\ \mu(q)),\ \max(\gamma(p),\ \gamma(q))\rangle,$$

$$V(p \veebar q) = \langle\max(\mu(p),\ \mu(q)),\ \min(\gamma(p),\ \gamma(q))\rangle,$$

$$V(p \supset q) = \langle\max(\gamma(p),\ \mu(q)),\ \min(\mu(p),\ \gamma(q))\rangle.$$

A given propositional form A (c. f. [7]: each proposition is a propositional form; if A is a propositional form, then $\neg$A is a propositional form; if A and B are propositional forms, then A & B, A $\veebar$ B, A $\supset$ B are propositional forms) will be called a <u>tautology</u>, <u>iff</u>

$$V(A) = \langle 1,\ 0\rangle,$$

for all valuation functions V, and an <u>intuitionistic fuzzy tautology</u>, <u>iff</u>

"if $V(A) = \langle a,\ b\rangle$, then $a \geq b$".

There exist some forms of an intuitionistic fuzzy Modus Ponens (see [3, 5]). The definition for the quantifiers is as follows:

$$V(\forall xA) = \langle\min_x \mu(A),\ \max_x \gamma(A)\rangle$$

and

$$V(\exists xA) = \langle\max_x \mu(A),\ \min_x \gamma(A)\rangle.$$

For the proposition p for which

$$V(p) = \langle a,\ b\rangle,$$

in [4] are defined the following operators:

$$V(\Box p) = \langle a,\ 1-a\rangle,$$

$$V(\Diamond p) = \langle 1-b, b \rangle,$$

which are analogous to the modal operators "necessity" and "possibility".

REFERENCES:

[1] K. Atanassov, Intuitionistic fuzzy sets, Fuzzy Sets and Systems, Vol. 20 (1986), No. 1, 87-96.

[2] K. Atanassov, More on intuitionistic fuzzy sets. Fuzzy Sets and Systems, 33, 1989, No. 1, 37-46.

[3] K. Atanassov, Two variants of intuitonistic fuzzy propositional calculus. Preprint IM-MFAIS-5-88, Sofia, 1988.

[4] K. Atanassov, Two variants of intuitionistic fuzzy modal logic. Preprint IM-MFAIS-3-89, Sofia, 1989.

[5] K. Atanassov, G. Gargov, Intuitionistic fuzzy logic. Compt. rend. Acad. bulg. Sci., Tome 43, N. 3, 1990, 9-12.

[6] A. Kaufmann, Introduction a la theorie des sous-ensembles flous, Paris, Masson, 1977.

[7] Mendelson E., Introduction to mathematical logic, Princeton, NJ: D. Van Nostrand, 1964.

LIST OF AUTHORS

Alexander Georgiev - Institute for Microsystems, Sofia

Alexander Savov - Institute for Microsystems, Sofia

Antoaneta Kirova - High Economical Institute, Sofia

Antonia Dimitrova - Institute for Microsystems, Sofia

Bistra Nikolova - Institute for Microsystems

Borjana Jordanova - Faculty of Mathematics, Kl. Ochridsky Univ., Sofia

Evgeni Tzolov - Institute for Microsystems, Sofia

Ewgeni Dimitrov - High Transport School, Sofia

Ilya Kazalarsky - Institute for Microsystems, Sofia

Ivan Hristozov - Institute for Microsystems, Sofia

Joseph Sorsich - 2-nd City Hospital, Sofia

Katja Stefanova - Institute for Standardization and Sertification, Sofia

Krassimir Atanassov - Institute for Microsystems, Sofia

Lilija Atanassova - 105 Al. Dimitrov School, Sofia

Ljubomir Hadjyisky - Technical University, Sofia

Ludmila Dimitrova - Institute of Chemical Engineering, Burgas

Maria Stefanova-Pavlova - Institute for Microsystems, Sofia

Martin Tetev - Faculty of Mathematics, Kl. Ochridsky Univ., Sofia

Niko Pehlivanov - Institute for Microsystems, Sofia

Pavlin Gyurov - Faculty of Mathematics, Kl. Ochridsky Univ., Sofia

Peter Georgiev - Faculty of Mathematics, Kl. Ochridsky Univ., Sofia

Plamen Petkov - NEFTOCHIM Petrochemical Combine, Bourgas

Radosvet Todotov - Centre for Scientific Information of Bulg. Acad. of Sciences, Sofia

Rossen Petrov - Institute for Microsystems, Sofia

Rumen Christov - Institute for Microsystems, Sofia

Sergej Nedev - Technical University, Sofia

Stanislav Pishmanov - Faculty of Mathematics, Kl. Ochridsky Univ., Sofia

Stefan Stefanov - Institute for Microsystems, Sofia

Stela Dimitrova - NEFTOCHIM Petrochemical Combine, Bourgas

Stojan Mihov - Faculty of Mathematics, Kl. Ochridsky Univ., Sofia

Stoian Garbov - Central Laboratory of Automation, Sofia

Trajana Kolarova - NEFTOCHIM Petrochemical Combine, Bourgas

www.ingramcontent.com/pod-product-compliance
Lightning Source LLC
LaVergne TN
LVHW060543180926
848329LV00007B/40